Semiconductor Memory Devices for Hardware-Driven Neuromorphic Systems

Semiconductor Memory Devices for Hardware-Driven Neuromorphic Systems

Editor

Seongjae Cho

MDPI • Basel • Beijing • Wuhan • Barcelona • Belgrade • Manchester • Tokyo • Cluj • Tianjin

Editor
Seongjae Cho
Department of Electronic
Engineering
Gachon University
Seongnam
Korea, South

Editorial Office
MDPI
St. Alban-Anlage 66
4052 Basel, Switzerland

This is a reprint of articles from the Special Issue published online in the open access journal *Electronics* (ISSN 2079-9292) (available at: www.mdpi.com/journal/electronics/special_issues/hdns).

For citation purposes, cite each article independently as indicated on the article page online and as indicated below:

LastName, A.A.; LastName, B.B.; LastName, C.C. Article Title. *Journal Name* **Year**, *Volume Number*, Page Range.

ISBN 978-3-0365-1734-6 (Hbk)
ISBN 978-3-0365-1733-9 (PDF)

Contents

About the Editor

Seongjae Cho

Seongjae Cho received B.S. and Ph.D. degrees in Electrical Engineering from Seoul National University, Seoul, Korea, in 2004 and 2010, respectively. He worked as an Exchange Researcher at the National Institute of Advanced Industrial Science and Technology (AIST), Tsukuba, Japan, in 2009, where he worked on three- and four-terminal FinFETs on 32-nm and 22-nm technology nodes.. Dr. Cho worked as a Postdoctoral Researcher at Seoul National University in 2010, where he developed three-dimensional nanoscale flash memory and devices and array architecture, and advanced low-power logic devices. From 2010 to 2013, Dr. Cho worked as a Postdoctoral Researcher at Stanford University, where he worked on photonic devices and circuits, heterostructure low-power electron devices, and biosensors with an emphasis on group-IV alloys. In 2013, he joined the Department of Electronic Engineering, Gachon University, Seongnam, Korea, where he is currently working as an Associate Professor. His research interests include flash and emerging memory devices, nanoscale CMOS devices, group-IV photonic devices, neuromorphic devices and integrated circuits, and novel processing-in-memory (PIM) cells. He is a Life Member of the Institute of Electronics and Information Engineers of Korea (IEIE) and a Senior Member of IEEE. He was the recipient of IEIE Haedong Young Engineer Award in 2011 and Academic Award from IEIE Semiconductor Materials and Devices Group in 2017.

Preface to "Semiconductor Memory Devices for Hardware-Driven Neuromorphic Systems"

Artificial intelligence (AI) is a technological area that has been under development for half a century. It is a term familiar to many people. AI has been exquisitely shaped into machine learning, and more recently, deep neural networks. As its evolution progresses, AI technology is infiltrating our daily lives more profoundly. Although AI technology has predominantly grown in computer and software engineering so far, further developments can be made in a hardware sense for higher system energy efficiency and more portable end-user-friendly edge applications. In order to more effectively mimic "our way of thinking" , mathematical analogy and the hardware-sense realization should go together hand in hand. AI can be more specifically termed as neuromorphic when the mathematical/algorithmic essences are realized by AI-oriented, specially designed hardware components. Hardware-sense AI can appear in general-purpose processing units made of conventional transistors. Integration of a large number of processing units can eventually mimic our way of thinking and can perform better depending on area. However, a lack of real AI may be perceived if volume and energy consumption are not considered. In order to address this deficiency, renovations should be realized at the device level. More synapse-like electron devices in terms of integration density, completeness in realizing biological synaptic behaviors, and energy-efficient operations are considered to be central for next-generation neuromorphic chips. The most important distinguishable feature between conventional AI chips and advanced neuromorphic systems is energy efficiency. However, only recently has this revolutionary synaptic device technology been implemented with semiconductor memory devices, materials, and processing technologies, with the aim of device scaling, data storage and processing, and low-power operation capabilities. It is the right time to investigate how the neuromorphic system and its building component technologies are developing. It cannot be underestimated that neural networks representing mathematical frames, energy-efficient memory-based synaptic devices, neurons and relevant circuits need to accompany one-another in good balance and harmony for ultra-low-power and super-light neuromorphic systems. This book will help the readership understand the evolutionary direction of neuromorphic systems, which is made in more hardware-driven ways, and provides perspectives in the relevant fields.

I deeply thank all the authors who have contributed the research articles with the best recency and also would like to give my sincere gratitude to my colleages, collaborators, family members, and my lifetime advisor, Prof. Byung-Gook Park. Also, the support for the research on neuromorphic devices and systems by Nano Material Technology Development Program through the National Research Foundation of Korea (NRF) funded by the Ministry of Science and ICT (MSIT) (NRF-2016M3A7B4910348) is acknowledged.

Seongjae Cho
Editor

Article

A Silicon-Compatible Synaptic Transistor Capable of Multiple Synaptic Weights toward Energy-Efficient Neuromorphic Systems

Eunseon Yu [1], Seongjae Cho [2,*] and Byung-Gook Park [3,*]

1 Department of Electrical and Computer Engineering, Purdue University, West Lafayette, IN 47906, USA; yu966@purdue.edu
2 Department of Electronics Engineering, Gachon University, Seongnam-si, Gyeonggi-do 13120, Korea
3 Department of Electrical and Computer Engineering, Seoul National University, Seoul 08826, Korea
* Correspondence: felixcho@gachon.ac.kr (S.C.); bgpark@snu.ac.kr (B.-G.P.); Tel.: +031-750-8722 (S.C.); +02-880-7270 (B.-G.P.)

Received: 15 July 2019; Accepted: 26 September 2019; Published: 30 September 2019

Abstract: In order to resolve the issue of tremendous energy consumption in conventional artificial intelligence, hardware-based neuromorphic system is being actively studied. Although various synaptic devices for the system have been proposed, they have shown limits in terms of endurance, reliability, energy efficiency, and Si processing compatibility. In this work, we design a synaptic transistor with short-term and long-term plasticity, high density, high reliability and energy efficiency, and Si processing compatibility. The synaptic characteristics of the device are closely examined and validated through technology computer-aided design (TCAD) device simulation. Consequently, full synaptic functions with high energy efficiency have been realized.

Keywords: energy consumption; hardware-based neuromorphic system; synaptic device; Si processing compatibility; TCAD device simulation

---

1. Introduction

Conventional computer architectures are mostly based on von Neumann's architecture since modern computer systems have been represented by electronic delay storage automatic calculator (EDSAC)—since 1949. The architecture consists of two main parts of processing and memory units performing the processes in the series' manner through single instruction and single data. Due to the physically differentiated system architecture, memory bus has been considered to be a bottleneck in determining the system processing speed, which is getting even worse in these days when big data are more increasingly demanded. In order to overcome this limit in the von Neumann computer architecture parallel processing capability of the artificial intelligent, of parallel processing with tremendous amount of data, contributions have been dedicated by the software-based neural networks. Although unimaginably many kinds of tasks have been accomplished by the software-driven technology in the given hardware system, with great resemblance to the way the human brain works, there is much room for enhancement of energy efficiency, which is the incomparable essence of biological system.

As a solution for the energy consumption issue, spiking neural network (SNN) is considered as one of the powerful schemes inspired by the biological system, which requires fundamental hardware innovation with synaptic transistors and neuron circuits [1,2]. Intellectual functions in human brain are determined by the strength and accuracy in connectivity among neurons. In human brain, there are a few tens of quadrillions of synapses and, through the synapses, humans become able to recognize, calculate, memorize, and learn. Thus, for hardware-driven neuromorphic systems to achieve more human-brain like computing efficiency, the synaptic device is required to have high scalability, multi-level weight adjustability, large inference margin, strong tolerance, and ultra-low energy consumption. Moreover, in order to gain higher access to the chip-level production lowering time and cost barriers, the SNN should be realized on the Si platform being helped by the mature Si processing technology.

A number of synaptic devices have been proposed with memristors such as resistive-switching random-access memory (ReRAM) and phase-change random-access memory (PcRAM). They are considered to be good candidates for the electronic synapse owing to their high structure simplicity and volume scalability, mainly by their great geometrical resemblance to the two-terminal structure of the biological synapse and energy efficiency [3–6]. Although memristors have these advantages, there is still room for further improving the rather low endurance and reproducibility and for enhancing the completeness in realizing the biological synaptic functions. Moreover, some of the existing memristor devices are not in consideration of Si processing compatibility. The simple structure requires functional compensation by additional devices or circuits, which might cause increased overhead in the SNN architecture [7–9]. In this work, a novel synaptic device has been designed, which has SiGe quantum well (QW) and Si_3N_4 charge-trap layer to realize the short-term potentiation (STP) and long-term potentiation (LTP), respectively, and its synaptic operations have been validated through technology computer-aided design (TCAD) device simulation, Silvaco Atlas [10]. Although the designed synaptic device is in a more complicated structure with a larger number of terminals compared with the two-terminal synaptic devices, it is capable of complementing the aforementioned weak points of the memristors, with an emphasis on higher energy efficiency and Si processing compatible materials. While most of the memristor synaptic devices have shown energy consumption higher than 1 pJ [11], the largest energy consumption required for a potentiation event has been demonstrated to be 1.51 fJ.

2. Device Structure and Design Strategies

More detailed explanations on the operation principles of the synaptic device and the models used in the device simulation along with the related physics are provided as follows. Figure 1 shows the schematic of the proposed synaptic device which has a p^+ SiGe layer at the drain-side channel and a charge-trap layer on the channel.

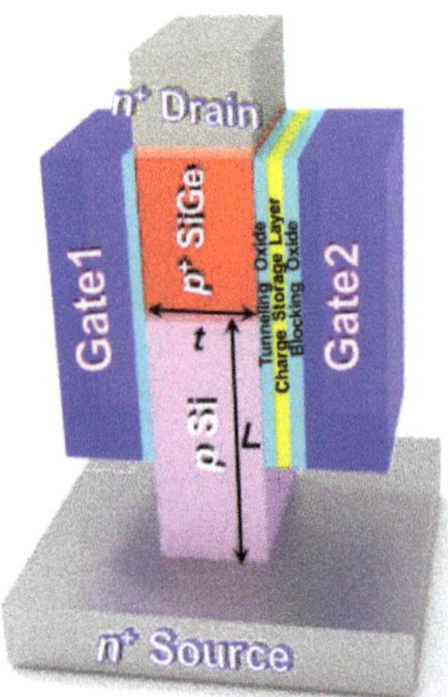

Figure 1. Schematic of the proposed synaptic device having an embedded SiGe quantum well and charge-storage layer for realizing the short-term and long-term plasticity, respectively.

As the number of potentiation pulses increases, the electrons in the SiGe valence band tunnel into the drain conduction band and fill the empty energy states. As a result, the holes generated in the SiGe layer are confined in the layer owing to a large valence-band offset (VBO) between Si and SiGe.

The confined holes give an effect of elevating the hole potential energy and the probability of hole tunneling into the nitride charge-trap layer, which realizes the LTP operation.

Designing a synaptic device with high reliability is paramount in building up a hardware architecture for the neuromorphic system. In order to demonstrate the device operation more accurately, multiple models are simultaneously activated. The mathematical and physical backgrounds of the used models can be glanced as follows. One of the essential differential equations used in the TCAD simulation is Poisson equation in Equation (1).

$$div(\varepsilon \nabla \psi) = -\rho \tag{1}$$

Here, ε is the local electrical permittivity of the material, ψ is the electrostatic potential, and ρ is the volume charge density.

$$\frac{\partial n}{\partial t} - \frac{1}{q} div \overrightarrow{J_{\rm n}} = G_{\rm n} - R_{\rm n} \tag{2}$$

Continuity equation in Equation (2) can be applied for obtaining the electron and hole current densities. n, $J_{\rm n}$, $G_{\rm n}$, and $R_{\rm n}$ are concentration of mobile electrons, areal electron current density, generation rate of electron, and recombination rate of electron, respectively. n can be substituted with p for hole description. q is the magnitude of electron charge. Based on the above equations, various models are equipped for higher accuracy and reliable simulation results. For an inversion layer mobility model, Lombardi model was used, which is suitable to non-planar devices, with dependences on both parallel and vertical electric fields, doping concentration, and temperature. The underlying physics comes from Matthiessen's rule.

$$\mu_{\rm T}^{-1} = \mu_{\rm AC}^{-1} + \mu_{\rm b}^{-1} + \mu_{\rm sr}^{-1} \tag{3}$$

Here, $\mu_{\rm T}$, $\mu_{\rm AC}$, $\mu_{\rm b}$, and $\mu_{\rm sr}$ indicate the total mobility, the surface mobility limited by scattering with acoustic phonons, the mobility limited by scattering with optical intervalley phonons, and the surface roughness factor, respectively.

$$f(E) = \frac{1}{1 + \exp\left(\frac{E}{kT_{\rm L}}\right)} \tag{4}$$

For carrier statistics, Fermi–Dirac statistics was employed. In Equation (4), $f(E)$ is the probability that an available electron state with energy E is occupied by an electron, k is Boltzmann constant, and $T_{\rm L}$ is lattice temperature. Moreover, the model is useful for the proposed device to describe the STP-to-LTP transition. The accumulated holes in the SiGe quantum well, which should be at the Fermi distribution tail, have higher probabilities of injection into the charge-trap layer. Moreover, non-local band-to-band tunneling calculation method was adopted, which has higher accuracy than the several tunneling models given as default in the TCAD simulation. This is due to the fact that the proposed device has the degenerately doped SiGe channel and drain, and the method calculates the tunneling probabilities by considering not only both forward and reverse tunneling currents but also the spatial variation of energy band and generation/recombination rates as shown in Equations (5) and (6).

$$J(E) = \frac{q}{\pi\hbar} \iint T(E)[f_{\rm l}(E + E_{\rm T}) - f_{\rm r}(E + E_{\rm T})]\rho(E_{\rm T})dEdE_{\rm T} \tag{5}$$

$$T(E) = \exp\left(-2\int_{x_{\rm start}}^{x_{\rm end}} k(x)dx\right) \tag{6}$$

Here, $J(E)$ is the net current density for a carrier with longitudinal (E) and transverse energy ($E_{\rm T}$) under the assumption that the tunneling current is the result of bidirectional transfers of carriers across the junction. $f_{\rm l}$ and $f_{\rm r}$ are the Fermi–Dirac functions using the quasi-Fermi levels in the left-side and right-side materials of the respective junctions. $\rho(E_{\rm T})$ and $k(x)$ represent the density of states corresponding to the transverse wavevector components and the wavevector at x. $T(E)$ indicates the tunneling probability for a carrier having an energy of E from the Wentzel–Kramers–Brilluoin

(WKB) approximation. Moreover, Shockley–Read–Hall recombination model, impact-ionization model, and bandgap narrowing model have been used. The aforementioned models are reflected for all the regions, and the non-local band-to-band model was applied locally between SiGe and Si where the tunneling events actually take place. In order to demonstrate the charge-trapping mechanism of nitride, a macro model (DYNASONOS) was employed, which includes various transport mechanisms such as thermionic emission, Poole–Frenkel emission, direct tunneling model, Fowler–Nordheim (FN) tunneling, and hot carrier injection at the same time. These models for the gate current are automatically applied for the Si_3N_4 layer and the regions in contact, which substantially affects the dynamics of the carriers moving into and out of the charge-trap layer. Without just using the default values given in the TCAD simulation package, the mobilities (μ) [12,13], saturation velocities (v_{sat}) [14–19], bandgap energy (E_g) [20], and electron affinity (χ) [21–27] of Si and SiGe have been fed into the device simulation [28]. This is because the SiGe layer, which stores holes, is considered as the important region for the synaptic operation. The values of the parameters are tabulated in Table 1.

Table 1. Parameters used in this work for Si and SiGe.

	μ_n [cm/V·s]	μ_p [cm/V·s]	$v_{sat,n}$ [cm/s]	$v_{sat,p}$ [cm/s]	χ [eV]	E_g [eV]
Si	1590.0	570.00	1.02×10^7	7.33×10^6	4.050	1.10
$Si_{0.7}Ge_{0.3}$	170.02	178.81	6.08×10^6	5.17×10^6	3.975	0.965

The SiGe layer is 50 nm long in the vertical direction and 50 nm wide (channel thickness = 50 nm). The *p*-type Si region is 100 nm long and the physical gate length (L_g) is 100 nm. Thus, whole SiGe region and the half of Si region are brought under the gate. In order to confine the holes generated over the potentiation process in the SiGe quantum well (QW) effectively, Ge fraction should be optimally controlled for a large valence-band offset (VBO) and the Si/SiGe interface status in the epitaxy processing as well, which is fixed to 0.3 throughout the design work. The gate oxide thickness for the gate 1 is 3 nm. The storage node is made up of oxide/nitride/oxide = 2/4/6 nm between the channel and the gate 2. The doping concentrations of source and drain junctions are both n^+-type 10^{20} cm^{-3}, and those of p^+ SiGe QW and *p*-type Si channel are 10^{18} cm^{-3} and 10^{16} cm^{-3}, respectively.

3. Design Results and Discussion

3.1. Design of Synaptic Device

In designing the synaptic device, the focus was placed on successfully emulating biological neural system with Si compatibility, high scalability, high reliability, and high energy efficiency. In order to meet the requirements, various approaches were performed including embedding SiGe layer. There is a large difference in E_g between Si and Ge and small difference in electron affinity (χ) so that most of the difference in energy bandgaps is transferred to VBO, which forms a hole QW in the SiGe region. Furthermore, SiGe is not only helpful in implementing potentiation mechanism but also in large current ratio between different weight states because its smaller E_g has the effect of lowering the potentiation voltage compared with the all-Si case. Employing these features of SiGe, the SiGe layer can be used as short-term storage node, making the device more energy-efficient.

Figure 2a shows the block diagram schematically explaining the learning rule of human brain by Hebbian's law [29]. Hebbian's law effectively dictates the correlation-based plasticity in the biological nervous system where the connectivity between pre-neuron and post-neuron, i.e., the synaptic conductance is strengthened by repeated firing events of the pre-neuron. An increased number of pulses in a given time, or equivalently, an increased pulse frequency enhances the transition probability of the synaptic device from short-term to long-term memory. Figure 2b,c shows the energy-band diagrams in the channel direction and metal-oxide-semiconductor direction from gate 1 to gate 2, respectively. For the potentiation operation, BTBT is adopted as the primary mechanism considering device reliability, scalability, and energy efficiency (Figure 2b). As shown in Figure 2b, for a potentiation pulse, the valence-band electrons in the SiGe QW can see the empty states of the conduction band of the Si drain junction. As the result, holes are generated and effectively confined in the SiGe layer due to the large VBO between SiGe and Si. The locally confined holes by QW VBO give an effect of elevating the QW potential and increasing the channel conductance temporarily [30], which corresponds to the STP. Then, if the potentiation pulses are repeatedly applied to the transistor before the generated holes are annihilated by either recombination or diffusion, i.e., if the holes are accumulated and their amount exceeds a certain threshold value in the SiGe QW, LTP is introduced. The accumulated holes with the energies at the Fermi-Dirac distribution tail have higher probabilities of injection into the nitride charge-trap layer. Once the holes are trapped in the nitride layer, they do not vanish for long time, which establishes the LTP function. Moreover, work functions of those two gates are optimally adjusted to locate the BTBT site not in the vicinity of the right-side channel in order to prevent a soft potentiation and to store the generated holes at the right-side of the channel, which leads to a stable and reproducible LTP operation as shown in Figure 2c. By reflecting the aforementioned approaches, design of a synaptic device meeting the requirements is realized.

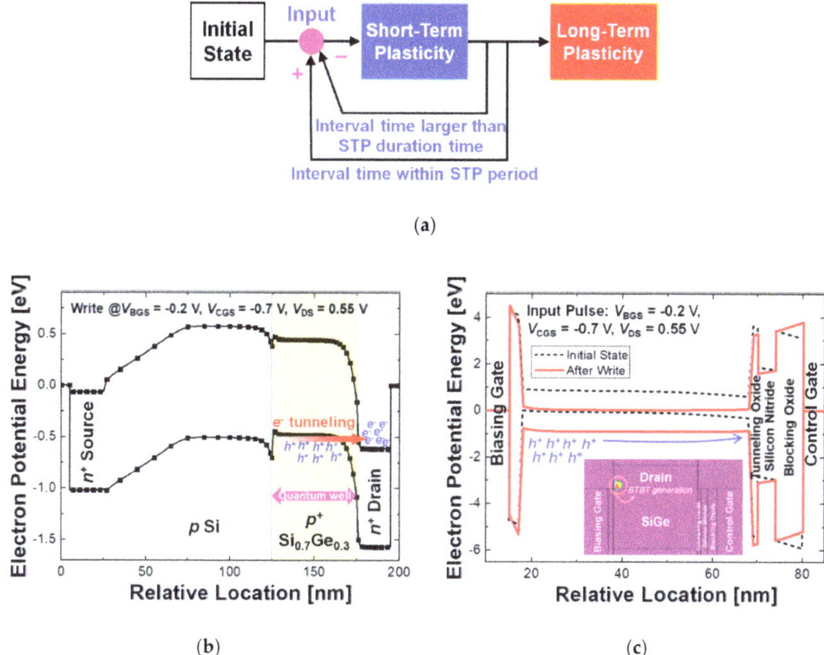

Figure 2. Operation principles of the synaptic device. (**a**) Hebbian's learning rule. (**b**) Energy-band diagram in the channel direction under the potentiation condition. (**c**) Energy-band diagram at the initial state and after potentiation state. The inset shows the band-to-band tunneling rate over a potentiation event.

3.2. Validation of Short- and Long-Term Plasticities

The proposed device has strong advantages particularly in energy-efficiency. There are many resources to make the device energy-efficient, such as introduction of SiGe QW, band-to-band tunneling mechanism, and STP characteristics. STP helps the device discriminate less important signals. Otherwise, when the weight of a synaptic device is changed at every input signal, the overall current over the synaptic device array would increase and large energy consumption is resulted.

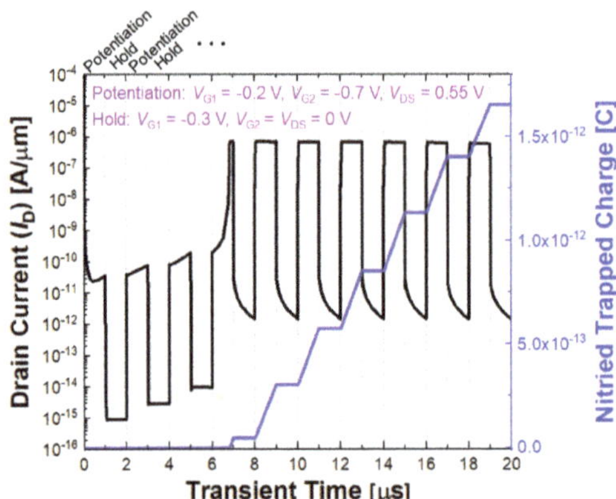

Figure 3. Drain current and nitride-trapped charges vs. learning pulses. The training pulse has a 1-μs width and a 1-μs interval.

Figure 3 indicates the timing diagram of drain current (I_D) and the amount of nitride-trapped charges as a function of time. The potentiation pulse is a set of (gate 1 voltage (V_{G1}), gate 2 voltage (V_{G2}), drain voltage (V_{DS}) = (−0.2 V, −0.7 V, 0.55 V), and the hold bias is (V_{G1}, V_{G2}, V_{DS}) = (−0.3 V, 0 V, 0 V). When a potentiation pulse is applied, holes are generated by band-to-band tunneling and confined in the SiGe layer. At the fourth pulse, I_D rapidly increases since the number of holes in the SiGe exceeds a certain threshold value and induces a drastic injection into the nitride charge-trap layer as shown in Figure 3. The trapped holes lower the threshold voltage of the synaptic device and increase the channel conductance.

Figure 4a shows the conduction-band edges obtained after different number of pulses are applied: 0, 1, 5, 10, 20, 30, 40, and 50 pulses. The insets depict the three-dimensional (3-D) contours of conduction band edge surfaces at the initial state and at a state after 30 pulses are applied, respectively. The line spectra representing the conduction band edges have been extracted from the channel vicinity of V_{G2} where the main current conduction path is formed. It is revealed that most of potential barrier lowering takes place by the holes in the SiGe region. Figure 4b plots the electron current density contours at the inference operations after different number of potentiation pulses: 1, 5, and 30 pulses. The inference process in the biological nervous system is analogous to the read operation in the memory array, and the electrical disturbance of the current data should be avoided. For the nondestructive inference, a voltage scheme was found to be V_{GS1} = V_{DS} = −0.1 V. As the number of pulses increases, more holes are populated in the charge-trap layer, and the potential barrier seen by the source electrons is lowered. Consequently, higher I_D is read at the same inference voltage as can be confirmed by Figure 4b.

Figure 5a demonstrates the transient characteristics of the synaptic transistor after different number of potentiation pulses. Through Figure 5a, it is confirmed that the proposed synaptic device is capable of both STP and LTP functions. The STP increases the channel conductivity for a short time, and the effect is diminished as time passes. As a result, I_D is eventually converged to the initial-state current level: The starting point can be varied but the final I_D is the same in the STP operation. On the other hand, I_D higher than the initial low current is consistently retained for up to 10^4 sec or more

When the synaptic device is brought into the LTP states. Here, it is notable that a large current difference takes place between states as the number of potentiation pulses increases. In Figure 5b, the actual transfer curves of the synaptic device obtained after the corresponding different number of pulses are applied in Figure 5a are depicted. In the STP operation, there is steady-state threshold voltage (V_{th}) shift. Once the device is in the LTP condition, a larger number of pulses lead to lower V_{th} without a temporal change. This is because the trapped holes in the nitride layer result in the inversion layer under the gate 2 at the inference bias. In Figure 5b, the proposed device demonstrates the large current ratio between high and low conductance states, which can be a beneficial aspect of a fully-Si electron device. The successfully suppressed leakage current stems from the high potential barrier constructed by the large VBO. If there is only STP, there would be no V_{th} shift. Only in the LTP condition, V_{th} begins the left-shifts due to the holes trapped in the nitride layer. It is shown that V_{th} of the proposed synaptic device is shifted by 1.5 V after 40 potentiation pulses are applied.

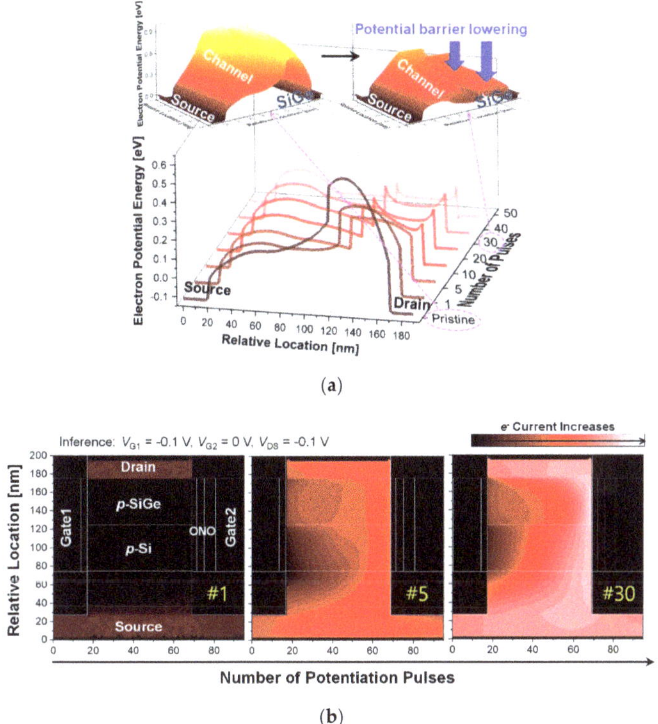

Figure 4. Analysis on potentiation operation. (**a**) Change in the conduction band surface with regard to the number of potentiation pulses: initial (left) and after 30 pulses (right). Line traces of the conduction band edges in the vicinity of gate 2. (**b**) Electron current densities after 1, 5, and 30 potentiation pulses applied to the synaptic device.

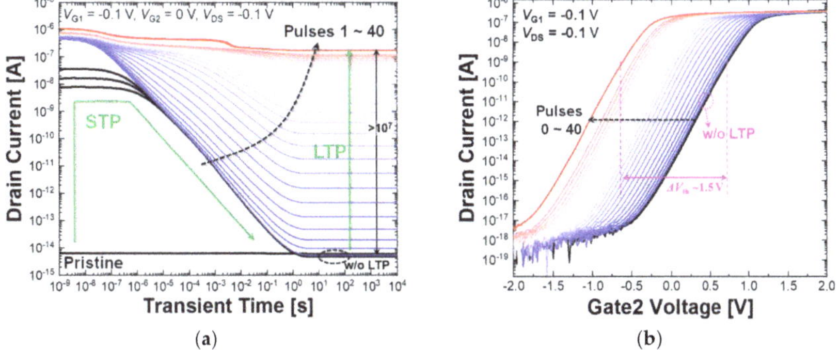

Figure 5. Electrical characteristics of the proposed synaptic device after different number of pulses: from 0 to 50 pulses. (**a**) Transient characteristics under read bias condition. (**b**) Transfer characteristics.

3.3. Interval Time Effects on STP and LTP Characteristics

Figure 6a–c shows how the transition from STP to LTP is made. As shown in Figure 6a, increasing the interval time between potentiation pulses makes it difficult to get into the LTP state. The holes in the SiGe layer temporarily generated by the pulses vanish by recombination and diffusion, which does not provide the boosting effect in band-to-band tunneling into the charge-trap layer. With the interval time of 1 ms, the synaptic device is not allowed to move to the LTP states as shown in Figure 6a and confirmed by Figure 6b. Figure 6b demonstrates the transient and DC characteristics under different interval time conditions for the same total number of potentiation pulses of 10. It is assured that a short enough time interval allows the synaptic device to enter the LTP states and modulate the electrical conductivity for learning.

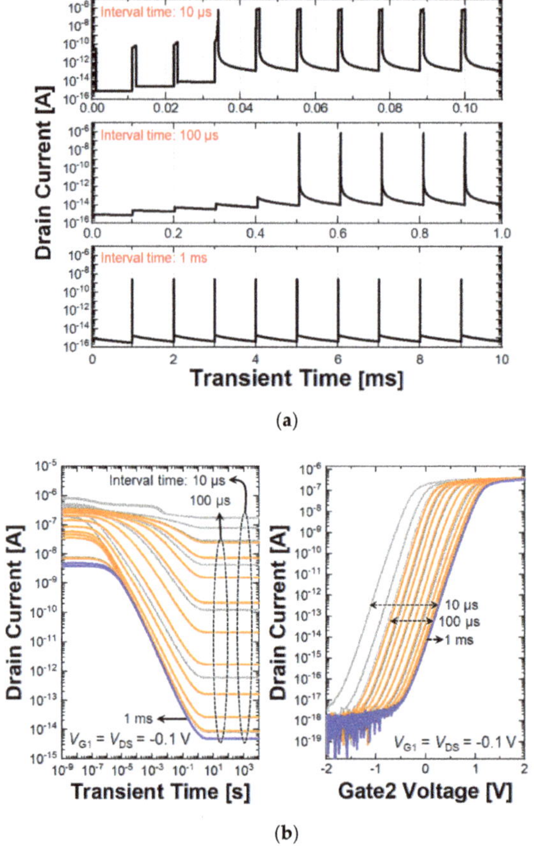

Figure 6. Operation characteristics depending on the number of pulse interval times. (**a**) Current changes with different pulse interval times: 10 μs, 100 μs, and 1 ms. (**b**) Transient (left) and DC sweep (right) characteristics with different interval times. For a shorter interval time at a given number of potentiation pulses, the saturation current increases and the V_{th} shift gets wider.

3.4. Spike-Timing-Difference Plasticity and Array Architecture

Table 2 summarizes the bias conditions for the synaptic operations and the required energy consumption per the realizable synaptic event along with the calculated synaptic device density. In order to exactly calculate the energy consumption, the current is integrated with time and multiplied by voltage. In Table 2, the energy consumption was obtained in the case of maximum value to consider the worst case. There are approximately 10^{15} synapses in the brain, about 1% of them are activated at the same time, and the frequency of neuron spike is about 10 Hz [31–33]. On this bases, human brain consumes power of ~20 W, and 55% out of the total power is consumed by the action potential [34–37]. In other words, it is assumed that power of 11 W is consumed for the synaptic activities. The power consumption per biological synaptic event is derived to be 1.1×10^{-12} W. Considering that each synaptic event has a ~100 ms duration, the energy consumption per synaptic event is calculated to be about 10 fJ. For the inference operation, 20 ns of rising and falling times and 10 ns of pulse duration are schemed, and then, this time period is multiplied by the inference voltage of −0.1 V and the current depending on weight for calculation of the energy consumption. The energies required for respective synaptic operations are summarized in Table 2, and most of them are very close to those for the biological synapse. Potentiation operation increases the conductance of the synaptic device so that a relatively large energy consumption is required; however, the amount is still low. Owing to the low operation voltage and tunneling-based injection mechanism, maximum energy consumption of only 0.52 pJ is needed for a potentiation event. For a depression event, the trapped holes tunnel back to the channel, which necessitates a relatively high operation voltage on the gate 2. However, even in the worst case, the required energy consumption is much lower than that for a potentiation event. Although the energy consumption for individual potentiation or depression event can be higher than that for an inference operation, the energy for an inference event can be spanned over a large range depending on the conductance. With the help of low-power and high-speed operation capabilities, the proposed synaptic transistor requires only a femto-joule energy even after 40 potentiation pulses are applied. Assuming that a unit cell has a footprint of 158 nm by 150 nm, the density of synaptic device array is calculated to be $9.09 \times 10^9/\text{cm}^2$. Here, the critical dimensions of the designed device and the metal pitches in one of the most recent memory technologies have been considered [38], where the wordline (WL) and bitline (BL) pitches are 48 and 54 nm, respectively.

Table 2. Energy consumption per realizable synaptic event and the calculated synapse density.

	V_{G1}	V_{G2}	V_{DS}	Time	Energy	
Potentiation	−0.2 V	−0.7 V	0.55 V	1 μs	0.52 fJ	
Depression	0 V	6 V	0 V	1 μs	1.51 fJ	
Inference	−0.1 V	0 V	−0.1 V	10 ns	Initial	6.42×10^{-24} J
					20 pulsed	1.87×10^{-16} J
					40 pulsed	5.24×10^{-16} J
Synapse density	$9.09 \times 10^9/\text{cm}^2$ (∵Unit cell size: 95 nm × 117 nm)					

Furthermore, the spike-timing-dependent plasticity (STDP) characteristics have been obtained by adjusting the pulse profile and time difference as demonstrated in Figure 7. The inset describes the potentiation and depression pulse schemes in the STDP simulations. The spiking pulse has 950 ns of rising time and 100 ns falling time, respectively, and the positive and negative voltage peak is 0.72 V in magnitude. The pre-neuron is connected with drain, and the post-neuron is with gate 1 and gate 2. When a pre-neuron spike comes earlier than a post-neuron one, the synaptic transistor is potentiated since the holes generated by tunneling operation from channel to drain are stored in the nitride charge-trap layer, which improves the conductance of the device. In the reversed order, the device is depressed owing to ejection of the trapped holes out of the nitride layer. When the timing difference between pre- and post-neuron spikes is larger than 900 ns, the conductance of the device is not changed but left as the initial value, which indicates that two neurons are not so closely correlated, and there is neither potentiation nor depression. It is noticeable that the synaptic transistor has a large current difference for the different spike timing, which substantially reduces the complexity of the sensing circuits and enhances the system reliability. Figure 8 demonstrates a presumable array architecture with the designed synaptic transistor for a hardware-driven neuromorphic system.

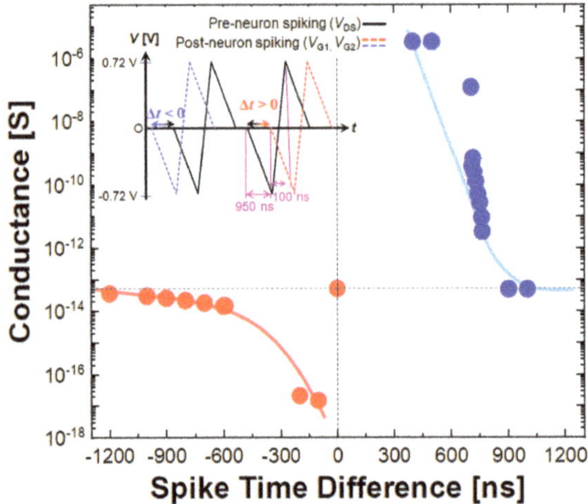

Figure 7. Spike-timing-dependent plasticity characteristic of the proposed synaptic device. The inset describes the potentiation and depression pulse schemes.

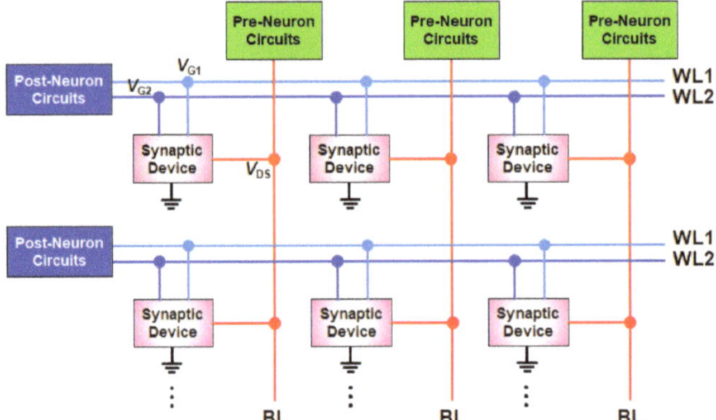

Figure 8. Array architecture with the proposed four-terminal synaptic transistor towards a high-density and high-reliability hardware-driven neuromorphic system.

4. Conclusions

In this work, a synaptic transistor having SiGe quantum well and nitride charge-trap layer was schemed and characterized by a series of rigorous simulation works. The synaptic device has successfully demonstrated the synaptic operations including STP, LTP, and inference with high energy efficiency not exceeding a two femto-joules in the worst case. Further, spike-timing-dependent plasticity was verified through a properly adjusted pulse scheme. A presumable array architecture is also conceived with the four-terminal synaptic device, and its density was calculated to be $9.09 \times 10^9/\text{cm}^2$ based on the interconnect schemes in the 18-nm DRAM technology node.

Author Contributions: E.Y. and S.C. conceived the device structure and wrote the manuscript. E.Y. performed the device simulations and evaluated the device scalability and array density. S.C. approved the simulation results in confirmation of the biological analogies. B.-G.P. conceived and developed the various types of hardware-driven neuromorphic systems, initiated the overall research project, and confirmed the validities of the simulated synaptic operations towards the artificial spike neural network.

Funding: This work was supported by Nano·Material Technology Development Program through the National Research Foundation of Korea (NRF) funded by the Ministry of Science and ICT (MSIT) (Grant

No. NRF-2016M3A7B4910348) and by Mid-Career Researcher Program through NRF funded by the MSIT (Grant No. NRF-2017R1A2B2011570).

Conflicts of Interest: The authors declare no conflict of interest.

References

1. Mead, C. Neuromorphic Electronic Systems. *Proc. IEEE* **1990**, *78*, 1629–1636. [CrossRef]
2. Burr, G.W.; Shelby, R.M.; Sebastian, A.; Kim, S.; Kim, S.; Sidler, S.; Virwani, K.; Ishii, M.; Narayana, P.; Fumarola, A.; et al. Neuromorphic computing using non-volatile memory. *Adv. Phys.* **2017**, *2*, 89–124. [CrossRef]
3. Shin, M.; Min, K.; Shim, H.; Kwon, Y. Investigation on Phase-change Synapse Devices for More Gradual Switching. *J. Semicond. Technol. Sci.* **2019**, *19*, 8–17. [CrossRef]
4. Pham, K.V.; Nguyen, T.V.; Tran, S.B.; Nam, H.; Lee, M.J.; Choi, B.J.; Truong, S.N.; Min, K.S. Memristor Binarized Neural Networks. *J. Semicond. Technol. Sci.* **2018**, *18*, 568–577. [CrossRef]
5. Sharad, M.; Fan, D.; Roy, K. Spin-neurons: A possible path to energy-efficient neuromorphic computers. *J. Appl. Phys.* **2013**, *114*, 234906. [CrossRef]
6. Cho, H.; Son, H.; Kim, J.-S.; Kim, B.; Park, H.-J.; Sim, J.-Y. Design of Digital CMOS Neuromorphic IC with Current-starved SRAM Synapse for Unsupervised Stochastic Learning. *J. Semicond. Technol. Sci.* **2018**, *18*, 65–77. [CrossRef]
7. Chu, M.; Kim, B.; Park, S.; Hwang, H.; Jeon, M.; Lee, B.H.; Lee, B.-G. Neuromorphic Hardware System for Visual Pattern Recognition with Memristor Array and CMOS Neuron. *IEEE Trans. Ind. Electron.* **2015**, *62*, 2410–2419. [CrossRef]
8. Kim, K.H.; Gaba, S.; Wheeler, D.; Cruz-Albrecht, J.M.; Hussain, T.; Srinivasa, N.; Lu, W. A Functional Hybrid Memristor Crossbar-Array/CMOS System for Data Storage and Neuromorphic Applications. *Nano Lett.* **2012**, *12*, 389–395. [CrossRef]
9. Park, S.; Noh, J.; Choo, M.; Sheri, A.M.; Chang, M.; Kim, Y.B.; Kim, C.J.; Jeon, M.; Lee, B.G.; Lee, B.H.; et al. Nanoscale RRAM-based synaptic electronics: Toward a neuromorphic computing device. *Nanotechnology* **2013**, *24*, 384009. [CrossRef]
10. SILVACO. *Atlas User's Manual*; SILVACO: Santa Clara, CA, USA, 2016.
11. Xu, W.; Min, S.Y.; Hwang, H.; Lee, T.W. Organic core-sheath nanowire artificial synapses with femtojoule energy consumption. *Sci. Adv.* **2016**, *2*, 1–7. [CrossRef]
12. Busch, G.; Vogt, O. Elektrische Leitfähigkeit und Halleffekt von Ge-Si-Legierungen. *Helv. Phys. Acta* **1960**, *33*, 437–458. [CrossRef]
13. Fischetti, M.V.; Laux, S.E. Band structure, deformation potentials, and carrier mobility in strained Si, Ge, and SiGe alloys. *J. Appl. Phys.* **1996**, *80*, 2234–2252. [CrossRef]
14. Ershov, M.; Ryahii, V. High-Field Electron Transport in SiGe Alloy. *Jpn. J. Appl. Phys.* **1994**, *33*, 1365–1372. [CrossRef]
15. Smith, J. Longitudinal Anisotropy of the High-Field Conductivity of *n*-Type Germanium at Room Temperature. *Phys. Rev.* **1969**, *178*, 1364–1367. [CrossRef]
16. Canali, C.; Majni, G.; Minder, R.; Ottaviani, G. Electron and Hole Drift Velocity Measurements in Silicon and Their Empirical Relation to Electric Field and Temperature. *IEEE Trans. Electron Devices* **1975**, *22*, 1045–1047. [CrossRef]
17. Ottaviani, G.; Reggiani, L.; Canali, C.; Nava, F.; Alberigi-Quaranta, A. Hole Drift Velocity in Silicon. *Phys. Rev. B* **1975**, *12*, 3318–3329. [CrossRef]
18. Ryder, E.J. Mobility of Holes and Electrons in High Electric Fields. *Phys. Rev.* **1953**, *90*, 766–769. [CrossRef]
19. Liou, T.S.; Wang, T.; Chang, C.Y. Analysis of High-Field Hole Transport Characteristics in $Si_{1-x}Ge_x$ Alloys with a Bond Orbital Band Structure. *J. Appl. Phys.* **1996**, *79*, 259–263. [CrossRef]
20. Braunstein, R.; Moore, A.R.; Herman, F. Intrinsic Optical Absorption in Germanium-Silicon Alloys. *Phys. Rev.* **1958**, *109*, 695–710. [CrossRef]
21. Chretien, O.; Apetz, R.; Souifi, A.; Vescan, L. $Si_{1-x}Ge_x$/Si valence band offset determination using current-voltage characteristics. *Thin Solid Films* **1997**, *294*, 198–200. [CrossRef]
22. People, R.; Bean, J.C.; Lang, D.V.; Sergent, A.M.; Störmer, H.L.; Wecht, K.W. Modulation doping in Ge_xSi_{1-x}/Si strained layer heterostructures. *Appl. Phys. Letts.* **1984**, *45*, 1231–1233. [CrossRef]

23. Takagi, S.; Hoyt, J.L.; Rim, K.; Welser, J.J.; Gibbons, J.F. Evaluation of the Valence Band Discontinuity of $Si/Si_{1-x}Ge_x/Si$ Heterostructures by Application of Admittance Spectroscopy to MOS Capacitors. *Electron Devices IEEE Trans.* **1998**, *45*, 494–501. [CrossRef]
24. Nauka, K.; Kamins, T.I.; Turner, J.E.; King, C.A.; Hoyt, J.L.; Gibbons, J.F. Admittance spectroscopy measurements of band offsets in $Si/Si_{1-x}Ge_x/Si$ heterostructures. *Appl. Phys. Letts.* **1992**, *60*, 195–197. [CrossRef]
25. Sant, S.; Lodha, S.; Ganguly, U.; Mahapatra, S.; Heinz, F.O.; Smith, L.; Moroz, V.; Ganguly, S. Band gap bowing and band offsets in relaxed and strained $Si_{1-x}Ge_x$ alloys by employing a new nonlinear interpolation scheme. *J. Appl. Phys.* **2013**, *113*, 033708. [CrossRef]
26. Walle, C.G.V.; Martin, R.M. Theoretical calculations of heterojunction discontinuities in the Si/Ge system. *Phys. Rev. B* **1986**, *34*, 5621–5634. [CrossRef]
27. Kurdi, M.; Sauvage, S.; Fishman, G.; Boucaud, P. Band-edge alignment of SiGe/Si quantum wells and SiGe/Si self-assembled islands. *Phys. Rev. B* **2006**, *73*, 195327. [CrossRef]
28. Yu, E.; Lee, W.J.; Jung, J.; Cho, S. Ultrathin SiGe Shell Channel *p*-Type FinFET on Bulk Si for Sub-10-nm Technology Nodes. *IEEE Trans. Electron Devices* **2018**, *65*, 1290–1297. [CrossRef]
29. Hebb, D.O. *The Organization of Behavior*; Wiley: New York, NY, USA, 1949.
30. Yu, E.; Cho, S.; Shin, H.; Park, B.G. A Band-Engineered One-Transistor DRAM with Improved Data Retention and Power Efficiency. *IEEE Electron Device Lett.* **2019**, *40*, 562–565. [CrossRef]
31. Kuzum, D.; Yu, S.; Wong, P. Synaptic electronics: Materials, devices and applications. *Nanotechnology* **2013**, *24*, 1–22. [CrossRef]
32. Huttenlocher, P. Synaptic density in human frontal cortex—Develop-metal changes and effects on aging. *Brain Res.* **1979**, *163*, 195–205. [CrossRef]
33. Lennie, P. The cost of cortical computation. *Curr. Biol.* **2003**, *13*, 493–497. [CrossRef]
34. Harris, J.; Jolivet, R.; Attwell, D. Synaptic Energy Use and Supply. *Neuron* **2012**, *75*, 762–777. [CrossRef] [PubMed]
35. Wang, Y.; Wang, R.; Xu, X. Neural Energy Supply-Consumption Properties Based on Hodgkin-Huxley Model. *Neural Plast.* **2017**, *2017*, 6207141. [CrossRef] [PubMed]
36. Attwell, D.; Laughlin, S. An Energy Budget for Signaling in the Gray Matter of the Brain. *J. Cereb. Blood Flow Metab.* **2001**, *21*, 1133–1145. [CrossRef]
37. Howarth, C.; Peppiatt-Wildman, C.; Attwell, D. The energy use associated with neural computation in the cerebellum. *J. Cereb. Blood Flow Metab.* **2010**, *30*, 403–414. [CrossRef]
38. Tech Insights. Samsung 18 nm DRAM Cell Integration: QPT and Higher Uniformed Capacitor High-k Dielectrics. Available online: https://www.techinsights.com/blog/samsung-18-nm-dram-cell-integration-qpt-and-higher-uniformed-capacitor-high-k-dielectrics (accessed on 14 June 2017).

Article

Solving Overlapping Pattern Issues in On-Chip Learning of Bio-Inspired Neuromorphic System with Synaptic Transistors

Hyungjin Kim [1,*] and Byung-Gook Park [2,*]

1 Department of Electronic Engineering, Yeungnam University, Gyeongsan 38541, Korea
2 Inter-University Semiconductor Research Center and Department of Electrical and Computer Engineering, Seoul National University, Seoul 08826, Korea
* Correspondence: hyungjin@yu.ac.kr (H.K.); bgpark@snu.ac.kr (B.-G.P.)

Received: 23 November 2019; Accepted: 18 December 2019; Published: 21 December 2019

Abstract: Recently, bio-inspired neuromorphic systems have been attracting widespread interest thanks to their energy-efficiency compared to conventional von Neumann architecture computing systems. Previously, we reported a silicon synaptic transistor with an asymmetric dual-gate structure for the direct connection between synaptic devices and neuron circuits. In this study, we study a hardware-based spiking neural network for pattern recognition using a binary modified National Institute of Standards and Technology (MNIST) dataset with a device model. A total of three systems were compared with regard to learning methods, and it was confirmed that the feature extraction of each pattern is the most crucial factor to avoiding overlapping pattern issues and obtaining a high pattern classification ability.

Keywords: neuromorphic system; on-chip learning; overlapping pattern issue; pattern recognition; synaptic device; spiking neural network

---

1. Introduction

Even though computing systems based on von Neumann architecture still dominate computer architecture, this architecture is considered inefficient for dealing with big data in the training of deep neural networks (DNNs) because of its serial signal processing [1]; therefore, a totally new computing system is required for the next generation of artificial intelligence. Recently, a bio-inspired neuromorphic system based on a spiking neural network (SNN) has been widely investigated because of its power-efficiency and parallel signal processing properties [2–5]. With regard to its application, the neuromorphic system, which is a hardware implementation of an artificial neural network, has been utilized mostly for pattern recognition [6–10], but also as a denoising auto encoder [11], for color image reconstruction [12], and for speech recognition [13]. In addition, various kinds of electronic devices have been studied as an artificial synaptic device, a crucial building block for constructing neuromorphic systems, including resistive switching materials [14–17], phase change materials [18–20], ferroelectric materials [21,22], and transistors [23–25]. Among them, transistor-based synaptic devices are considered as having better reliability characteristics and device variation for very-large-scale integration (VLSI) implementation of neural networks compared to their counterparts.

In our previous works, we reported a synaptic transistor with an asymmetric dual-gate structure as having short- and long-term memories and spike-timing dependent plasticity (STDP) characteristics [26–28], and its fabrication method [29]. In this work, a system-level study of a SNN for pattern recognition is presented with a binary modified National Institute of Standards and Technology (MNIST) handwritten dataset. The necessity of an inhibitory synaptic component is analyzed in order to solve an overlapping pattern issue when it comes to pattern recognition for on-chip learning of bio-inspired neuromorphic systems in the form of SNNs.

2. Device Model of Synaptic Transistor for System-Level Study

A schematic view of weight modulation in the synaptic transistor is illustrated in Figure 1a. As the pre-synaptic spikes are applied to the first gate (G1) and the drain, excess holes are generated by impact ionization and accumulate in the floating body region. The impact generation region expands as a result of the positive feedback between impact generation rate and accumulated holes. Afterwards, newly generated hot carriers near the second gate (G2) are injected into the nitride layer depending on the second gate voltage (V_{G2}). The device is potentiated and depressed when holes and electrons are stored in the nitride layer because of the threshold voltage (V_T) change. These weight modulation characteristics of the synaptic transistor are incorporated into a device model with a voltage-controlled current source (VCCS) [30] based on the gate current caused by hot carrier injection [31], as shown in Figure 1b. The VCCS delivers the second gate current (I_{G2}) to the nitride layer, which is modeled by the gate current flowing by hot carrier injection as a function of V_{G2} as per the following equation:

$$I_{G2} \text{ represented by VCCS} = \alpha \cdot (V_{G1} - V_T)^2 \cdot {V_{G2}}^2 \cdot \exp(-1/V_{G2}) \quad (1)$$

where α is a fitting coefficient. The type and number of injected carriers are determined depending on V_{G2} so that the amount of V_T change (ΔV_T) per each pre- and post-synaptic spike is calculated, providing good agreement with the measured data [28].

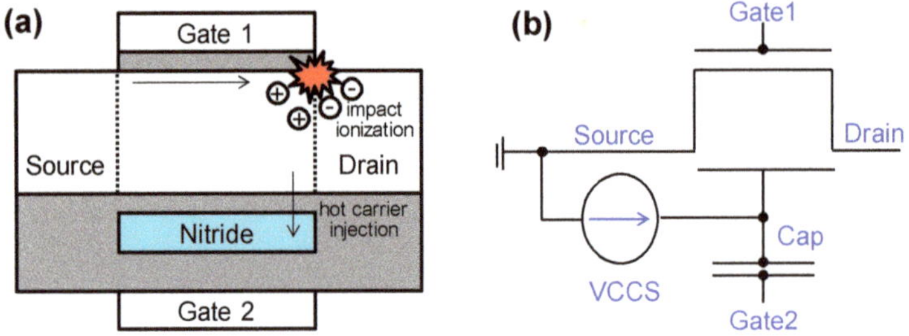

Figure 1. (**a**) Schematic view of weight modulation in the synaptic transistor with dual-gate structure. (**b**) Device model of the synaptic transistor with a voltage-controlled current source (VCCS).

3. Results and Discussion

With the help of the developed device model, the performance of the SNN composed of the synaptic transistors was studied with regard to pattern recognition. A 784 × 10 single-layer SNN was constructed to train and test 28 × 28 binary MNIST images (60,000 training images and 10,000 testing images). A total of 784 synaptic transistors were connected to each output node as shown in Figure 2. Charges were integrated at a capacitor node while pre-synaptic spikes were applied to each synaptic transistor, and a post-synaptic neuron circuit generated post-synaptic spikes at the output node when the node voltage of the capacitor exceeded V_T of the neuron circuit [32]. The spike generation rate of each post-synaptic neuron circuit was considered as the intensity of the output node; therefore, the system was considered successful in pattern recognition when the answer node fired most among all the output nodes during test operation. The reason why recognition accuracy was calculated in this manner is that the weight sum of transferred currents (I_E) to the output node, which is the most congruous to the testing sample, was expected to be the largest owing to the potentiated synaptic transistors in the shape of the digit, leading to high current flows.

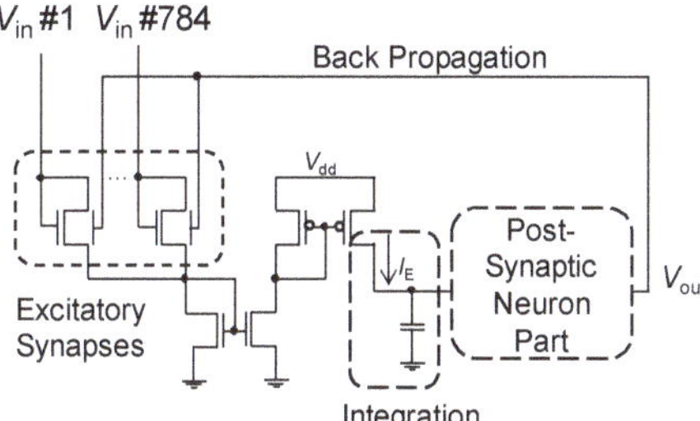

Figure 2. Single-layer spiking neural network (SNN) for pattern recognition with synaptic transistors and neuron circuit.

Figure 3a shows how the system was trained using the binary MNIST images and STDP characteristics. The pre-synaptic spikes were applied to the corresponding synaptic transistors with different timing depending on their colors: black with Δt = 0.5 μs and white with Δt = −0.5 μs, compared to a teaching signal which was given to the output node matching to the digit of the training sample. Therefore, V_T was increased (depression) for the synaptic transistors representing black pixels (background) and decreased (potentiation) for white pixels (handwritten digit). Figure 3b shows the classification rate of the SNN with untrained testing samples as a function of the number of trained samples.

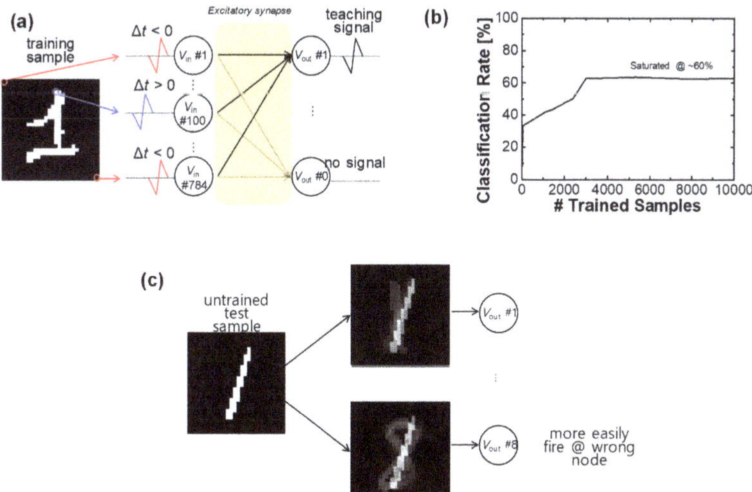

Figure 3. (**a**) Learning method using the spike-timing dependent plasticity (STDP) rule. (**b**) Classification rate depending on the number of trained samples. (**c**) Overlapping pattern issue.

The accuracy rate became rapidly saturated due to the nonlinear weight modulation characteristics coming from the hot carrier injection model. The more electrons or holes were trapped in the nitride layer, the less likely were additional electrons or holes to be injected due to the potential inhibition by the already stored ones. The saturated accuracy rate of over 3000 trained samples was about 60%, which is quite low compared to other SNN systems because of the overlapping pattern issue. Figure 3c describes how the overlapping pattern issue degrades the classification rate. The output nodes having more white pixels in their weight maps, such as eight or zero, have a higher probability to fire, even though they do not match the digits of test samples, leading to a low recognition rate of the ones that have less white pixels (such as digit 1).

Figure 4a compares two weight maps in the form of ΔV_T at the same scale, which were learned through the STDP rule and transferred from an artificial neural network (ANN) through off-chip learning; here, the synaptic weights of the ANN were converted to the ones of the SNN to be proportional to their square roots, so that the transferred I_E can be in line with the weight sum of the ANN with a rectified linear unit (ReLU), which is one of the most popular activation functions in ANNs because of the lack of vanishing gradients problems compared to other ones, such as sigmoid or a hyperbolic tangent [33–35]. The former looks like carving digits to the synaptic devices, whereas the latter is well characterized by the features of each digit. That is why the hardware-based SNN has a poor accuracy of 60% because its weight map does not reflect the characteristics of each digit. In the case of the STDP method, the V_T is modulated only according to whether a training sample is the answer or not, and the amount of V_T change is determined by the amount of already stored carriers in the nitride layer. However, the amount of V_T change is adjusted in the case of the ANN according to a backpropagation algorithm. Illustrated in Figure 4b are the transformation processes of the weight maps for digit 8 as the training progressed for the two cases. The carved pattern on the weight map by the STDP method becomes clearer in the direction in which it can fire frequently by digit input samples; however, the transferred weight map from the ANN exhibits its unique features in fine detail, so that all the weight maps have higher classification accuracies, which means that even a narrower memory window of the synaptic transistors can provide a higher accuracy when the weight map reflects the unique characteristics of the training images which are supposed to be classified. Figure 4c plots the classification rates for each digit depending on the methods. The poor accuracies, especially digit 1 and digit 9, have been highly improved by adopting the transferred synaptic weights, leading to 87.6% of the total accuracy. In addition, the most noteworthy thing is that the classification rates of the transferring method and the ANN itself are almost the same for every single digit. It is believed that the SNN using the transferred weight maps and the ANN with ReLU are equivalent in their operations in the respect that the intensity of the output nodes can correspond to the firing rate [36].

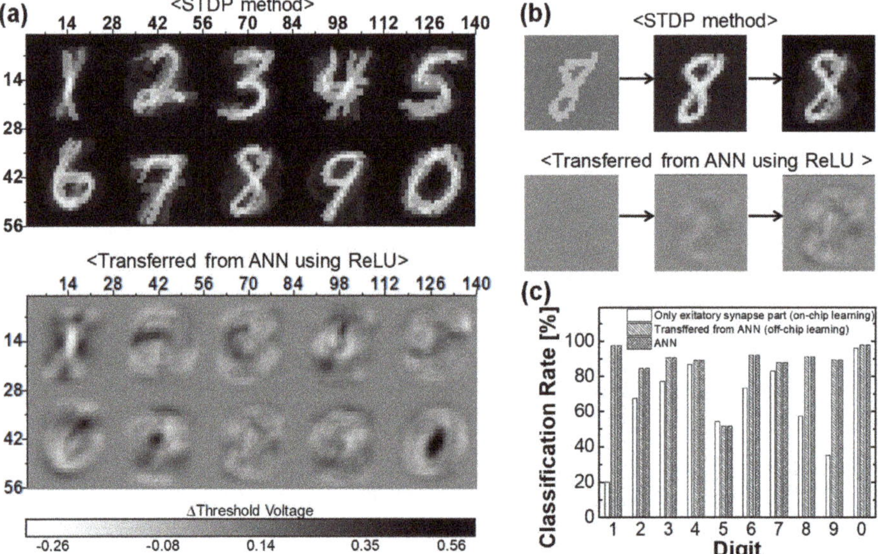

Figure 4. Transferred synaptic weights from artificial neural networks. (**a**) Comparison of the weight maps learned by the STDP method and transferred from an artificial neural network (ANN). (**b**) Training progress of each weight map. (**c**) Comparison of the classification accuracy of each digit for the STDP method, transferred synaptic weights method, and ANN.

In order to reduce the classification error caused by the overlapping pattern issue discussed above, the inhibitory synaptic devices with the same weight maps as the excitatory ones are added as shown in Figure 5a. As in the previous method, the input signals are applied to the excitatory synapses corresponding to their own pixels in the case of the white pixels; at the same time, the input signals are applied to the inhibitory synapses in the case of the black pixels. This change in the manner of classification leads to the result that if the testing samples cover not only their own digits but also other digits, the remaining parts contribute to a subtraction of the weight sum by the current flows (I_I) through the inhibitory synaptic transistors as shown in Figure 5b. The overlapping pattern issue can be significantly solved in this way because it mainly comes from the contribution of the remaining parts to undesired firings.

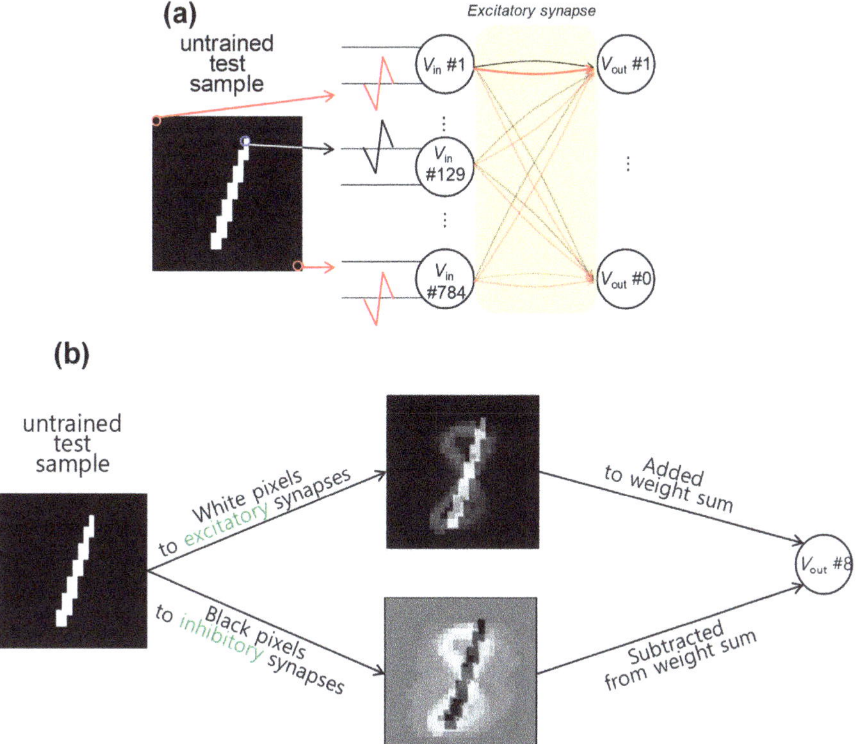

Figure 5. (**a**) Illustration of the classification with the addition of inhibitory synaptic transistors. (**b**) How the inhibitory synaptic transistors solve the overlapping pattern issue.

Figure 6a shows the accuracies as a function of the number of training samples for various ratios between the channel widths of the excitatory synaptic transistors (W_{ex}) and the inhibitory ones (W_{in}). The accuracy is improved by 10% at W_{in}/W_{ex} = 0.1; however, it starts decreasing after that and reaches the bottom (nearly 0% instead of 10%) when W_{in}/W_{ex} = 0.5. This is because the output nodes cannot fire when W_{in} is too wide. The number of black pixels is larger than that of white pixels and I_I is higher than I_E in most testing samples when W_{in}/W_{ex} exceeds 0.5. Figure 6b compares the classification rate of each digit for those two SNN systems. It is noteworthy that the accuracies of the digits which have a small number of white pixels, such as one, is significantly enhanced from 19 to 60%, while the accuracies of other digits maintain similar values. It is confirmed that the addition of an inhibitory synapse part can effectively solve the misclassified cases stemming from the overlapping pattern problem.

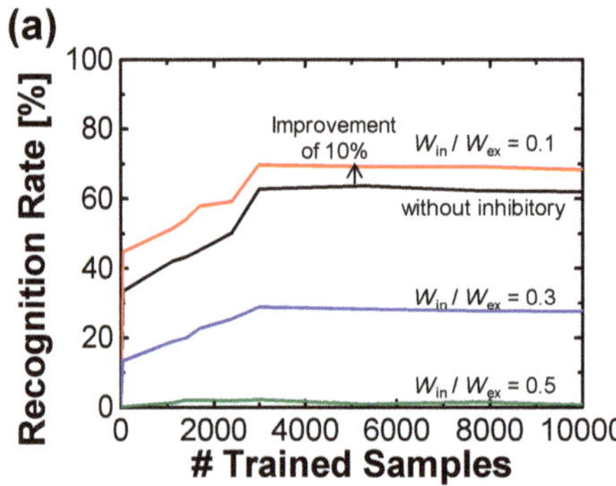

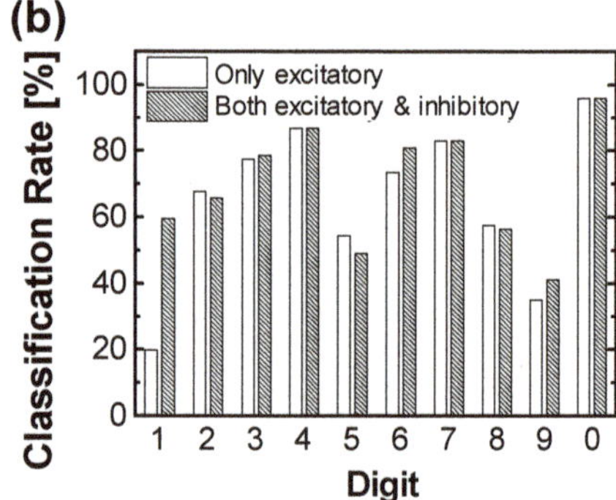

Figure 6. (**a**) Classification rates after adding the inhibitory synapse part. (**b**) Classification accuracy of each digit depending on the learning method and system structure.

4. Conclusions

In conclusion, we presented a system-level study regarding pattern recognition with the help of a device model. The device model was developed with a VCCS based on measured data and gate current by hot carrier injection. A total of three SNN systems were constructed and analyzed using binary MNIST images. A SNN with only the excitatory synaptic transistors trained under the STDP rule had a poor classification rate with 60% of the total accuracy because of the pattern overlapping issue. This dramatically improved to 87.6% in the case of a SNN with transferred synaptic weights from an ANN using ReLU. The difference between those two systems was whether the region representing the unique features of each digit was potentiated or the handwritten digit region was just carved. The addition of inhibitory synaptic transistors with the same weight maps improved the classification accuracy by 10% by solving the overlapping pattern problem, which comes from the fact that the output nodes having more white pixels tend to fire to unmatched training samples. These results lead us to conclude that these SNN systems and learning methods provide a framework for future studies about hardware-based neuromorphic systems using both excitatory and inhibitory synaptic devices for pattern recognition applications.

Author Contributions: H.K. and B.-G.P. conceived the device structure and modeling and wrote the manuscript. All authors have read and agreed to the published version of the manuscript.

Funding: This work was supported in part by the NRF funded by the Korean government under Grant 2019M3F3A1A03079821, 2019R1I1A3A01061262, 2019R1A6C1030008 and in part by the Nano-Material Technology Development Program through the NRF funded by the Ministry of Science, ICT and Future Planning (2016M3A7B4910348).

Conflicts of Interest: The authors declare no conflict of interest.

References

1. Backus, J. Can programming be liberated from the von Neumann style?: A functional style and its algebra of programs. *Commun. ACM* **1978**, *21*, 613–641. [CrossRef]
2. Jeong, D.S.; Kim, K.M.; Kim, S.; Choi, B.J.; Hwang, C.S. Memristors for Energy-Efficient New Computing Paradigms. *Adv. Electron. Mater.* **2016**, *2*, 160090. [CrossRef]
3. Merolla, P.A.; Arthur, J.V.; Alvarez-Icaza, R.; Cassidy, A.S.; Sawada, J.; Akopyan, F.; Jackson, B.L.; Imam, N.; Guo, C.; Nakamura, Y.; et al. A million spiking-neuron integrated circuit with a scalable communication network and interface. *Science* **2014**, *345*, 668–673. [CrossRef] [PubMed]
4. Misra, J.; Saha, I. Artificial neural networks in hardware: A survey of two decades of progress. *Neurocomputing* **2010**, *74*, 239–255. [CrossRef]

5. Maass, W. Networks of spiking neurons: The third generation of neural network models. *Neural Netw.* **1997**, *10*, 1659–1671. [CrossRef]
6. Cohen, E.; Malka, D.; Shemer, A.; Shahmoon, A.; Zalevsky, Z.; London, M. Neural networks within multi-core optic fibers. *Sci. Rep.* **2016**, *6*, 29080. [CrossRef] [PubMed]
7. Hwang, S.; Kim, H.; Park, J.; Kwon, M.-W.; Baek, M.-H.; Lee, J.-J.; Park, B.-G. System-level simulation of hardware spiking neural network based on synaptic transistors and I&F neuron circuits. *IEEE Electron Device Lett.* **2018**, *39*, 1441–1444.
8. Seo, M.; Kang, M.-H.; Jeon, S.-B.; Bae, H.; Hur, J.; Jang, B.C.; Yun, S.; Cho, S.; Kim, W.-K.; Kim, M.-S.; et al. First demonstration of a logic-process compatible junctionless ferroelectric finfet synapse for neuromorphic applications. *IEEE Electron Device Lett.* **2018**, *39*, 1445–1448. [CrossRef]
9. Park, Y.J.; Kwon, H.T.; Kim, B.; Lee, W.J.; Wee, D.H.; Choi, H.-S.; Park, B.-G.; Lee, J.-H.; Kim, Y. 3-D stacked synapse array based on charge-trap flash memory for implementation of deep neural networks. *IEEE Trans. Electron Devices* **2018**, *66*, 420–427. [CrossRef]
10. Sung, C.; Lim, S.; Kim, H.; Kim, T.; Moon, K.; Song, J.; Kim, J.-J.; Hwang, H. Effect of conductance linearity and multi-level cell characteristics of TaO_x-based synapse device on pattern recognition accuracy of neuromorphic system. *Nanotechnology* **2018**, *29*, 115203. [CrossRef]
11. Kim, H.; Hwang, S.; Park, J.; Yun, S.; Lee, J.-H.; Park, B.-G. Spiking neural network using synaptic transistors and neuron circuits for pattern recognition with noisy images. *IEEE Electron Device Lett.* **2018**, *39*, 630–633. [CrossRef]
12. Shabairou, N.; Cohen, E.; Wagner, O.; Malka, D.; Zalevsky, Z. Color image identification and reconstruction using artificial neural networks on multimode fiber images: Towards an all-optical design. *Opt. Lett.* **2018**, *43*, 5603–5606. [CrossRef] [PubMed]
13. Truong, S.N.; Ham, S.-J.; Min, K.-S. Neuromorphic crossbar circuit with nanoscale filamentary-switching binary memristors for speech recognition. *Nanoscale Res. Lett.* **2014**, *9*, 629. [CrossRef] [PubMed]
14. Prezioso, M.; Merrikh-Bayat, F.; Hoskins, B.; Adam, G.; Likharev, K.K.; Strukov, D.B. Training and operation of an integrated neuromorphic network based on metal-oxide memristors. *Nature* **2015**, *521*, 61–64. [CrossRef]
15. Ambrogio, S.; Balatti, S.; Milo, V.; Carboni, R.; Wang, Z.-Q.; Calderoni, A.; Ramaswamy, N.; Ielmini, D. Neuromorphic learning and recognition with one-transistor-one-resistor synapses and bistable metal oxide RRAM. *IEEE Trans. Electron Devices* **2016**, *63*, 1508–1515. [CrossRef]
16. Park, J.; Kwak, M.; Moon, K.; Woo, J.; Lee, D.; Hwang, H. TiOx-based RRAM synapse with 64-levels of conductance and symmetric conductance change by adopting a hybrid pulse scheme for neuromorphic computing. *IEEE Electron Device Lett.* **2016**, *37*, 1559–1562. [CrossRef]
17. Kim, S.; Kim, H.; Hwang, S.; Kim, M.-H.; Chang, Y.-F.; Park, B.-G. Analog synaptic behavior of a silicon nitride memristor. *ACS Appl. Mater. Interfaces* **2017**, *9*, 40420–40427. [CrossRef]
18. Kim, S.; Ishii, M.; Lewis, S.; Perri, T.; BrightSky, M.; Kim, W.; Jordan, R.; Burr, G.W.; Sosa, N.; Ray, A.; et al. NVM neuromorphic core with 64k-cell (256-by-256) phase change memory synaptic array with on-chip neuron circuits for continuous in-situ learning. In Proceedings of the 2012 International Electron Devices Meeting, San Francisco, CA, USA, 7–9 December 2012; pp. 443–446.
19. Tuma, T.; Le-Gallo, M.; Sebastian, A.; Eleftheriou, E. detecting correlations using phase-change neurons and synapses. *IEEE Electron Device Lett.* **2016**, *37*, 1238–1241. [CrossRef]
20. Kuzum, D.; Jeyasingh, R.G.; Lee, B.; Wong, H.-S.P. Nanoelectronic programmable synapses based on phase change materials for brain-inspired computing. *Nano Lett.* **2011**, *12*, 2179–2186. [CrossRef]
21. Oh, S.; Kim, T.; Kwak, M.; Song, J.; Woo, J.; Jeon, S.; Yoo, I.K.; Hwang, H. $HfZrO_x$-based ferroelectric synapse device with 32 levels of conductance states for neuromorphic applications. *IEEE Electron Device Lett.* **2017**, *38*, 732–735. [CrossRef]
22. Mulaosmanovic, H.; Ocker, J.; Müller, S.; Noack, M.; Müller, J.; Polakowski, P.; Mikolajick, T.; Slesazeck, S. Novel ferroelectric FET based synapse for neuromorphic systems. In Proceedings of the 2017 Symposium on VLSI Technology, Kyoto, Japan, 5–8 June 2017; pp. T176–T177.
23. Wang, J.; Li, Y.; Liang, R.; Zhang, Y.; Mao, W.; Yang, Y.; Ren, T.-L. Synaptic computation demonstrated in a two-synapse network based on top-gate electric-double-layer synaptic transistors. *IEEE Electron Device Lett.* **2017**, *38*, 1496–1499. [CrossRef]
24. Wan, X.; Yang, Y.; Feng, P.; Shi, Y.; Wan, Q. Short-term plasticity and synaptic filtering emulated in electrolyte-gated IGZO transistors. *IEEE Electron Device Lett.* **2016**, *37*, 299–302. [CrossRef]

25. Shi, J.; Ha, S.D.; Zhou, Y.; Schoofs, F.; Ramanathan, S. A correlated nickelate synaptic transistor. *Nat. Commun.* **2013**, *4*, 2676. [CrossRef] [PubMed]
26. Kim, H.; Park, J.; Kwon, M.-W.; Lee, J.-H.; Park, B.-G. Silicon-based floating-body synaptic transistor with frequency dependent short-and long-term memories. *IEEE Electron Device Lett.* **2016**, *37*, 249–252. [CrossRef]
27. Kim, H.; Cho, S.; Sun, M.-C.; Park, J.; Hwang, S.; Park, B.-G. Simulation study on silicon-based floating body synaptic transistor with short-and long-term memory functions and its spike timing-dependent plasticity. *J. Semicond. Technol. Sci.* **2016**, *16*, 657–663. [CrossRef]
28. Kim, H.; Hwang, S.; Park, J.; Park, B.-G. Silicon synaptic transistor for hardware-based spiking neural network and neuromorphic system. *Nanotechnology* **2017**, *28*, 405202. [CrossRef]
29. Kim, H.; Sun, M.-C.; Hwang, S.; Kim, H.-M.; Lee, J.-H.; Park, B.-G. Fabrication of asymmetric independent dual-gate FinFET using sidewall spacer patterning and CMP processes. *Microelectron. Eng.* **2018**, *185*, 29–34. [CrossRef]
30. Kim, S.; Lee, S.-H.; Kim, Y.-G.; Cho, S.; Park, B.-G. Highly compact and accurate circuit-level macro modeling of gate-all-around charge-trap flash memory. *Jpn. J. Appl. Phys.* **2016**, *56*, 014302. [CrossRef]
31. Sonoda, K.; Tanizawa, M.; Shimizu, S.; Araki, Y.; Kawai, S.; Ogura, T.; Kobayashi, S.; Ishikawa, K.; Eimori, T.; Inoue, Y.; et al. Compact modeling of a flash memory cell including substrate-bias-dependent hot-electron gate current. *IEEE Trans. Electron Devices* **2004**, *51*, 1726–1733. [CrossRef]
32. Park, J.; Kwon, M.-W.; Kim, H.; Hwang, S.; Lee, J.-J.; Park, B.-G. Compact neuromorphic system with four-terminal si-based synaptic devices for spiking neural networks. *IEEE Trans. Electron Devices* **2017**, *64*, 2438–2444. [CrossRef]
33. LeCun, Y.; Bengio, Y.; Hinton, G. Deep learning. *Nature* **2015**, *521*, 436–444. [CrossRef] [PubMed]
34. Nair, V.; Hinton, G.E. Rectified linear units improve restricted boltzmann machines. In Proceedings of the 27th International Conference on Machine Learning, Haifa, Israel, 21–24 June 2010; pp. 807–814.
35. Maas, A.L.; Hannun, A.Y.; Ng, A.Y. Rectifier nonlinearities improve neural network acoustic models. In Proceedings of the 30th International Conference on Machine Learning, Atlanta, GA, USA, 16–21 June 2013.
36. Diehl, P.U.; Neil, D.; Binas, J.; Cook, M.; Liu, S.-C.; Pfeiffer, M. Fast-classifying, high-accuracy spiking deep networks through weight and threshold balancing. In Proceedings of the 2015 International Joint Conference on Neural Networks, Killarney, Ireland, 12–17 July 2015; pp. 1–8.

Article

3-D Synapse Array Architecture Based on Charge-Trap Flash Memory for Neuromorphic Application

Hyun-Seok Choi [1], Yu Jeong Park [2], Jong-Ho Lee [3] and Yoon Kim [1,*]

1 School of Electrical and Computer Engineering, University of Seoul, Seoul 02504, Korea; cawai7@naver.com
2 Applied Materials Korea, Ltd., Hwaseong-si, Gyeonggi-do 18364, Korea; pyoojeng@naver.com
3 Department of Electrical and Computer Engineering, Seoul National University, Seoul 08826, Korea; jhl@snu.ac.kr
* Correspondence: yoonkim82@uos.ac.kr; Tel.: +82-02-6490-2352

Received: 29 November 2019; Accepted: 29 December 2019; Published: 30 December 2019

Abstract: In order to address a fundamental bottleneck of conventional digital computers, there is recently a tremendous upsurge of investigations on hardware-based neuromorphic systems. To emulate the functionalities of artificial neural networks, various synaptic devices and their 2-D cross-point array structures have been proposed. In our previous work, we proposed the 3-D synapse array architecture based on a charge-trap flash (CTF) memory. It has the advantages of high-density integration of 3-D stacking technology and excellent reliability characteristics of mature CTF device technology. This paper examines some issues of the 3-D synapse array architecture. Also, we propose an improved structure and programming method compared to the previous work. The synaptic characteristics of the proposed method are closely examined and validated through a technology computer-aided design (TCAD) device simulation and a system-level simulation for the pattern recognition task. The proposed technology will be the promising solution for high-performance and high-reliability of neuromorphic hardware systems.

Keywords: 3-D neuromorphic system; 3-D stacked synapse array; charge-trap flash synapse

1. Introduction

Neuromorphic systems have been attracting much attention for next-generation computing systems to overcome the von Neumann architecture [1–5]. The term "neuromorphic" refers to an artificial neural system that mimics neurons and synapses of the biological nervous system [3]. A neuron generates a spike when a membrane potential which is the result of the spatial and temporal summation of the signal received from the pre-neuron exceeds a threshold, and the generated spike is transmitted to the post-neuron. A synapse refers to the junction between neurons, and each synapse has its own synaptic weight which is the connection strength between neurons [6]. In a neuromorphic system, synaptic weight can be represented by the conductance of synapse device.

The requirements of a synapse device to implement a neuromorphic system are as follows: small cell size, low-energy consumption, multi-level operations, symmetric and linear weight change, high endurance and complementary metal-oxide semiconductor (CMOS) compatibility [5]. Various memory devices, such as static random-access memories (SRAM) [7], resistive random-access memories (RRAM) [8], phase change memories (PCM) [9], floating gate- memories (FG-memory) [10] and charge-trap flash memories [11] have been proposed to implement the synapse operation. Among them, charge-trap flash (CTF) devices have good CMOS compatibility and excellent reliability [12–15].

In our previous work, we proposed a 3-D stacked synapse array based on a charge trap flash (CTF) device [11]. Three-dimensional stacking technology is currently used in the commercialized Not AND

(NAND) flash memory products for ultra-high density [14]. Similarly, a 3-D stacked synapse array has the advantage of chip-size reduction when implementing very-large-size artificial neural networks. Consequently, it has the potential to be a promising technology for implementing neuromorphic hardware systems. For the design of the 3-D stacked synapse array architecture, there are several issues. At the full array level, how to operate each layer selectively and how to efficiently form the metal interconnects with peripheral circuits are critical issues. At the device level, how to implement accurate synaptic weight levels with low energy consumption is an important issue. Especially, linear and symmetric synaptic weight (conductance) modulations are essential to improve the accuracy of neuromorphic hardware systems [1–4].

In this paper, we examine these issues and suggest two improvements in terms of an architecture design and a device operation method. The rest of the paper is structured as follows: Section 2 contains design methods based on the viewpoint of a full-chip architecture. In this section, we review the 3-D stacked synapse array structure developed in the previous work [11] and propose an improved version of the 3-D stacked synapse array architecture to solve the unwanted problem of the previous version. In Section 3, we propose an improved programming method to obtain linear and symmetric conductance changes. Using a pattern recognition application with the Modified National Institute of Standards and Technology (MNIST) database, we demonstrate the improvement of the proposed method.

2. Design Methods of 3-D Synapse Array Architecture

In general, a large-size artificial neural network that has a large number of synaptic weights and neuron layers is required to obtain high performance artificial intelligence tasks. In the case of the ImageNet classification challenge, state-of-the-art deep neural network (DNN) architectures have 5~155M synaptic weight parameters [16]. In order to implement efficiently a large-size artificial neural network on a limited-size hardware chip, we proposed the 3-D stacked synapse array structure (Figure 1) in the previous work [11].

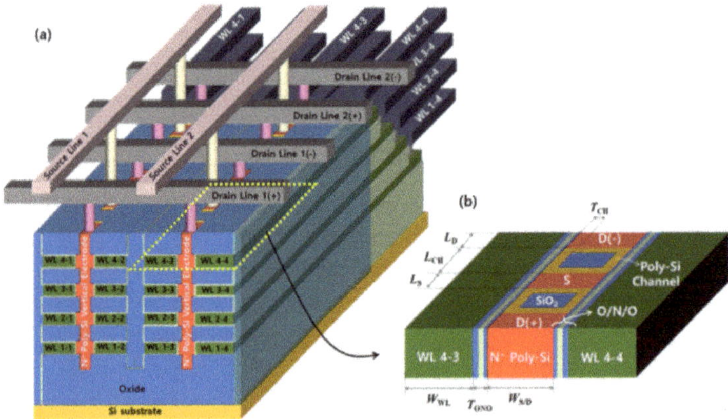

Figure 1. 3-D synapse array structure [11]. (**a**) 3-D stacked synapse device; (**b**) Unit synapse cell structure.

Unit synapse cell is composed of two CTF devices having two drain nodes (D(+), D(−)) and common source node(S). The D(+) part is connected to the output neuron circuit to increase membrane potential, acting as an excitatory synaptic behavior. The D(−) part is connected to the output neuron circuit to decrease membrane potential, acting as an inhibitory synaptic behavior. By using this configuration, it can be represented the negative and positive weight at the same time. As summarized in Table 1, the CTF device has several advantages over other non-volatile memory devices. First, it does not need an additional selector device because the three-terminal MOSFET-based unit cell has a built-in selection operation. Second, it has perfect CMOS compatibility. Third, the linear and incremental modulation of the weight (conductance) can be more easily achieved because its conductance is determined by the number of trapped charges. Fourth, it has good retention reliability characteristics. On the other hand, the drawback of CTF is large power consumption during program operation. Therefore, CTF devices are the best solution for off-chip learning-based neuromorphic systems where frequent weight updates do not occur.

Table 1. Comparison between non-volatile memory devices for neuromorphic hardware systems.

	RRAM	PCM	STT-MRAM	CTF
Device Structure	2 terminals	2 terminals	2 terminals	3 terminals
Selector	needed	needed	needed	unneeded
Cell Size	4 ~ 12 F^2	4 ~ 12 F^2	6 ~ 20 F^2	4 ~ 8 F^2
CMOS Compatibility	good	good	moderate	very good
Multi-Level Operation	good	good	moderate	very good
Weight Change	abrupt SET	abrupt RESET	stochastic change	good symmetric
Write Latency	20 ~ 100 ns	40 ~ 150 ns	2 ~ 20 ns	>1 μs
Write Energy	low	mid	low	mid~high
Retention	moderate	good	good	very good

The proposed 3-D stacked synapse array structure is based on the word-line stacking method which is similar to the commercialized V-NAND flash memory. Therefore, it has the advantage of utilizing the existing stable process methods used in V-NAND flash memory.

A key issue in the design of 3-D stacked synapse array architecture is the metal interconnection. For example, a 4-layer stacked synapse array would have four times as many word lines as a 2-D synapse array. If the word-line (WL) decoder is connected by a conventional metal interconnection method, the vertical length of the WL decoder (HWL_Decoder) will increase as illustrated in Figure 2, resulting in an enormous loss of area efficiency in terms of full-chip level architecture. To solve this issue, we proposed the smart design of a layer select decoder with 3-D metal line connection in the previous work [11]. As shown in Figure 3a, the area of WL decoder is not increased, and a layer select decoder is added to selectively operate each stacked layer. A layer select decoder delivers the gate voltages generated by the WL decoder to the WLs of the selected layer. It is important to note that the vertical length of a layer select decoder is the same as that of the WL decode, and the horizontal length is only 4 F×N where F is the minimum feature size and N is the number of staked layers. The specific structure of the transistors and metal interconnects is depicted in our previous paper [11].

The top-view layout of the 3-D synapse array architecture is illustrated in Figure 4. The layer select decoder is composed of pass transistors. The pass transistors are arranged next to each word line and are connected one-to-one with each WL contact. The gate nodes of the pass transistors are vertically connected to form a layer select line (LSL) that is controlled by LSL control circuit. Through this configuration, each stacked layer can be selectively operated while maintaining compact full-chip configuration efficiency. For example, if the turn-on voltage is applied to L4 and the turn-off voltages are applied to L1~L3, pass transistors corresponding to L = 4 are activated. Consequently, the WL voltages generated in the WL decoder are transferred to the forth-layer WLs (L = 4).

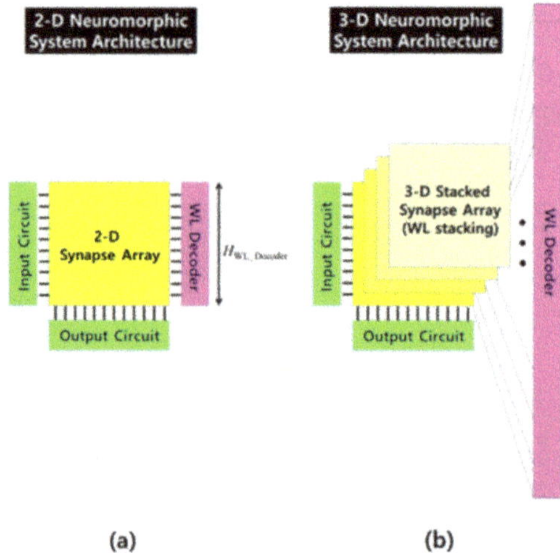

Figure 2. Metal interconnection scheme of synapse array architecture. (**a**) 2-D neuromorphic system architecture; (**b**) 3-D neuromorphic system architecture (a bad design example).

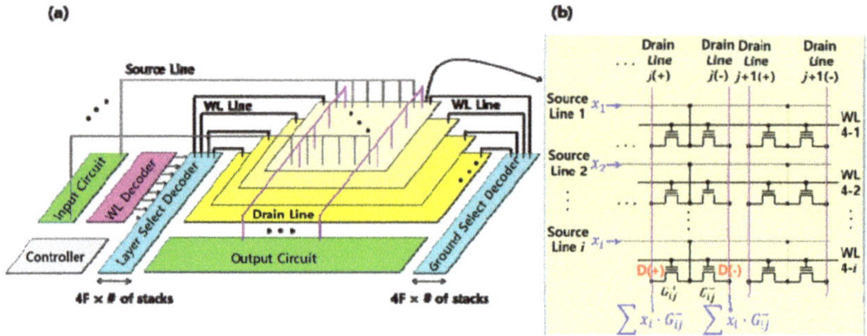

Figure 3. Schematic of the proposed 3-D synapse array architecture. (**a**) Metal interconnection of a full-chip architecture; (**b**) Each synapse layer configuration to implement artificial neural network.

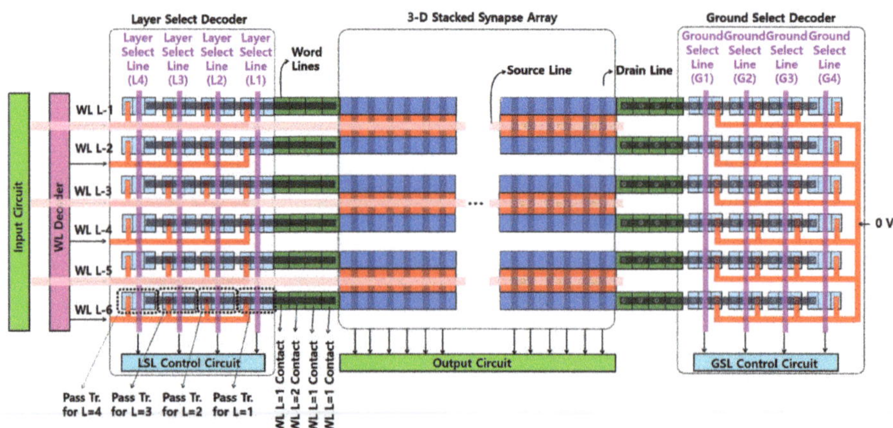

Figure 4. Top view image of the revised synapse array architecture.

In this paper, we proposed an improved architecture design compared to the previous work, adding the ground select decoder as shown in Figure 4. If there is only a layer select decoder, the WLs of the unselected stacked layer are on a floating state because they are not connected to the WL decoder. In this case, the potential of the WLs of the unselected layer varies due to the capacitive coupling between the stacked WLs. In the worst case, the WLs of unselected layers located above or below ($L = n - 1$ or $L = n + 1$) the selected layer ($L = n$) may be boosted together when a high voltage is applied to the selected WLs. To solve this inherent risk of the architecture of the previous version, a ground select decoder that applies a turn-off voltage (0 V) to the WLs of the unselected layer is added to the right side of the main 3-D stacked synapse array as shown in Figure 4.

The detailed manufacturing process of the 3-D synapse array was described in our previous paper [11]. The revised synapse array architecture can be made with the same process method. Since the newly added ground select decoder structure has the same structure as the layer select decoder, it can be made by just adding the same layout as the layer select decoder.

To validate the synaptic operations of the designed CTF-based synapse device, the technology computer-aided design (TCAD) device simulation (Synopsys Sentaurus [17]) was used. The specific device parameters are summarized in Table 2. Electrical characteristics of the designed synapse device are discussed in the next chapter.

Table 2. Physical parameters of the device used for electrical simulation.

	Value
$L_S = L_D$	50 nm
L_{CH}	100 nm
T_{CH}	10 nm
$T_{O/N/O}$	3/6/6 nm
$W_{WL} = W_{S/D}$	100 nm

3. Results

3.1. Synapse Device Operation

In the proposed synapse array (Figure 3b), synaptic weight (w_{ij}) of the artificial neural network is represented as follows:

$$w_{ij} = G^{+}{}_{ij} - G^{-}{}_{ij}. \quad (1)$$

As depicted in Figure 3b, $G^{+}{}_{ij}$ and $G^{-}{}_{ij}$ are the conductances of the D(+) CTF device and D(−) CTF device, respectively. Each conductance is determined by the amount of trapped charge in each charge-trap layer (silicon nitride). For the conductance modulation, hot-electron injection (HEI) and hot-hole injection (HHI) can be used as a charge injection mechanism. The potentiation process of increasing the synaptic weight can be performed by increasing $G^{+}{}_{ij}$ and decreasing $G^{-}{}_{ij}$. On the other hand, the depression process of decreasing the synaptic weight can be carried out by decreasing $G^{+}{}_{ij}$ and increasing $G^{-}{}_{ij}$. Using a technology computer-aided design (TCAD) device simulation (Synopsys Sentaurus), we verify two pulse schemes for the modulation of synaptic weight. A successive-pulse programming scheme and the incremental-step-pulse programming (ISPP) scheme are illustrated in Figure 5a,b, respectively.

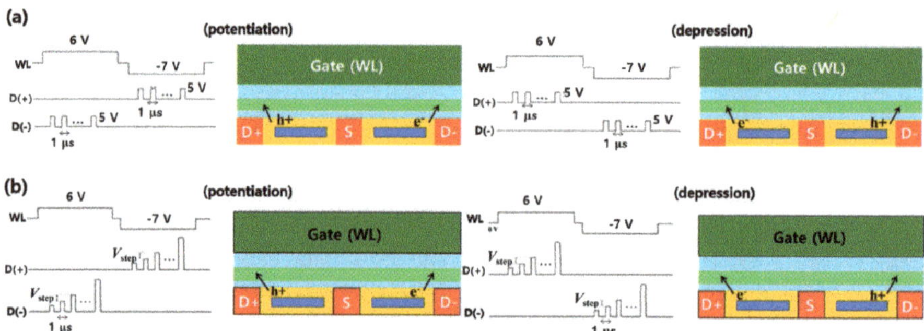

Figure 5. Programming schemes for synaptic weight (conductance). (**a**) Successive-pulse programming scheme; (**b**) Incremental-step-pulse programming scheme.

A successive-pulse programming is a method of continuously applying drain pulses with the same voltage as shown in Figure 5a. In this programming scheme, the amount of conductance change is controlled by the number of applied drain pulses. When the drain pulse is applied, the sign of the gate voltage determines whether HEI or HHI occurs. If the drain pulse is applied when the gate bias is positive (6 V), HEI occurs. In this case, the threshold voltage increases by the trapped electron and the conductance decreases. On the other hand, if the drain pulse is applied when the gate bias in negative (−7 V), HHI occurs. In this case, the threshold voltage decreases by the trapped hole and the conductance increases. The proposed unit synapse cell is composed of two CTF devices. Consequently, the potentiation operation is conducted simultaneously by HHI in the D(+) CTF device and HEI in the D(−) CTF device. The depression operation is conducted by HEI in the D(+) device and HHI in the D(−) device.

The ISPP is used for the program scheme of NAND flash memory [18]. The program pulse is increased by a constant value V_{step} after each program step, as shown in Figure 5b. In our previous paper, only successive-pulse programming was used. In this work, we applied the ISPP method to the conductance modulation of our designed synapse device. Using a TCAD device simulation, we compared the conductance modulation characteristics of successive-pulse programming and the ISPP. As shown in Figure 6, the ISPP scheme shows better synaptic behavior than the successive-pulse scheme. The ISPP scheme showed that the conductance changes linearly according to the number of applied pulses. Also, the range of available synaptic weights (memory window) can be further increased. Consequently, the ISPP scheme can adjust the synaptic weight more accurately than the successive-pulse programming scheme during the learning process. However, the ISPP scheme also has a drawback. In order to determine the start pulse voltage, a verify operation is required prior to programming to check the current synaptic weight value. Therefore, the ISPP scheme can increase the accuracy of the learning process, but also increases time and energy consumption.

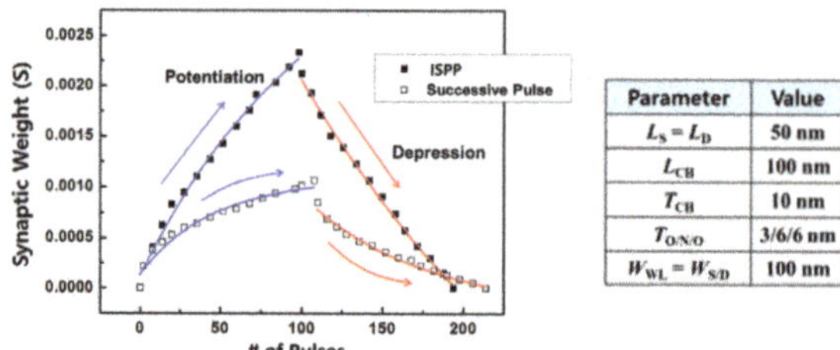

Parameter	Value
$L_S = L_D$	50 nm
L_{CB}	100 nm
T_{CB}	10 nm
$T_{O/N/O}$	3/6/6 nm
$W_{WL} = W_{S/D}$	100 nm

Figure 6. Gradual changes of synaptic weights by successive-pulse programming and incremental-step-pulse programming (ISPP).

3.2. System-Level Simulation for Pattern Recognition

To validate the functionality of the proposed programming schemes, the single-layer artificial neural network system for the Modified National Institute of Standards and Technology (MNIST) pattern recognition was simulated. The MNIST database is a large database of handwritten digits, which contains about 60,000 learning images and 10,000 test images [19]. A total of 784 input neurons represent 28 × 28 pixels of each image and 10 output neurons represent 10 digits (0 ~ 9). We also used a rectifier linear unit (ReLU) as an activation function of neuron, which is one of the popular activation functions [20]. For the learning process, a supervised learning method was used. At first, the error was calculated at the output neurons. Next, the target change in synaptic weight (the number of programming pulses) was determined by the gradient descent method. After that, the synaptic weight value is updated based on fitted equations for the conductance modulation characteristics of a successive-pulse programming scheme and the ISPP scheme.

Figure 7a shows the system-level simulation result of the pattern recognition accuracy with the 10,000 test image samples. Compared to our previous work [11], the ISPP scheme can increase recognition accuracy by about 6% (a successive-pulse programming scheme in our previous work: 79.83% [11], and the ISPP scheme in this work: 85.9%). This result is in good agreement with the other papers that the linear conductance modulation characteristic is essential for the better performance of neuromorphic systems [5,21]. The synaptic weight maps after training 10,000 samples with the ISPP scheme are illustrated in Figure 7b.

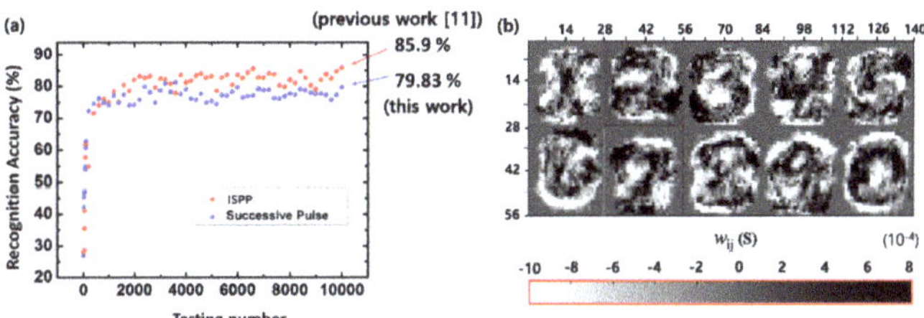

Figure 7. Modified National Institute of Standards and Technology (MNIST) pattern recognition result. (**a**) Recognition accuracy comparison between a successive-pulse programming and the ISPP; (**b**) Synaptic weight map after training 10,000 samples with the ISPP scheme.

In addition, we examined the synaptic weight modulation characteristics according to the various values of V_{step} in the ISPP scheme. As illustrated in Figure 8a, a smaller V_{step} allows for fine conductance modulation, which means that the number of the synaptic weight level can be increased. As a result, the fine conductance modulation ability with a smaller V_{step} can obtain more accurate pattern recognition rate, as shown in Figure 8b. It should be noted, however, that the retention characteristics (the ability to distinguish each level for a long time) can deteriorate when the interval between each synaptic weight level becomes narrow. Therefore, the magnitude of V_{step} should be determined considering the trade-off relationship between the retention characteristic and the accuracy.

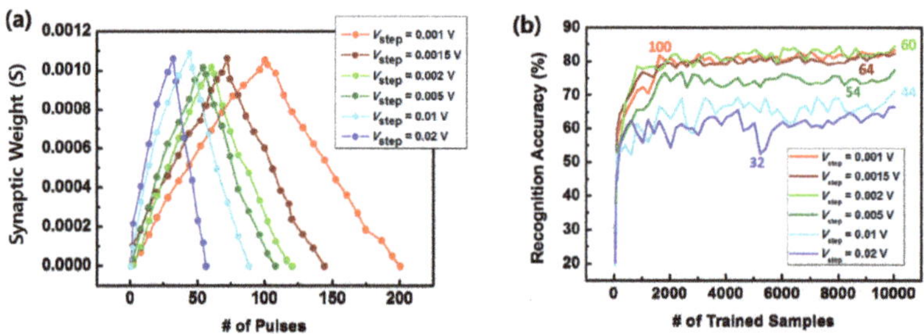

Figure 8. MNIST pattern recognition result by using the ISPP scheme. (**a**) The gradual conductance change by applying various V_{step}; (**b**) Recognition accuracy as a function of the training samples for the various V_{step}. The number means the weight levels (maximum pulse number).

4. Discussion

Currently, numerous researches based on different types of nonvolatile memory devices have been conducted to implement neuromorphic hardware systems. Table 3 summarizes some of the research results.

Table 3. Comparison between several research results of neuromorphic applications.

	This Work	Previous Work [11]	[22]	[23]	[24]
Synapse Device	CTF	CTF	CTF	RRAM	PRAM
Array Architecture	3-D array	3-D array	2-D array	2-D array	2-D array
Neuron Layer	single-layer	single-layer	single-layer	single-layer	multi-layer
Learning Type	supervised	supervised	supervised	supervised	unsupervised
Recognition Rate	85.9%	79.8%	84%	87.9%	95.5%
Result Type	simulation	simulation	measurement	measurement	simulation

Almost all previous studies are based on the 2-D synapse array structure, but for the first time we proposed the 3-D stacked synapse array structure. This paper has addressed several issues associated with the design of the 3-D synapse array architecture in terms of a full-chip level. This will be an important guideline for designing a 3-D stacked synapse array. The approach of stacking CTF devices is a mature technology that has been already used in commercialized 3-D NAND flash memories. Consequently, the proposed 3-D synapse architecture is expected to have a high possibility of actual mass production. Also, it can achieve excellent reliability by utilizing the various technologies used in NAND flash memory. For example, we have demonstrated that the ISPP method can improve the pattern recognition accuracy of a neuromorphic system.

For future work, we will demonstrate the superiority of the proposed 3-D synapse architecture based on an actual fabricated array. In addition, application researches to various artificial neural networks such as a convolutional neural network (CNN) and a recurrent neural network (RNN) will be a crucial topic.

5. Conclusions

We proposed a 3-D synapse array architecture based on a CTF memory device. To resolve the drawback of the previous version of the architecture, a ground select decoder was newly added. Also, we introduced the ISPP scheme to improve the linearity of the conductance modulation. The characteristics of synaptic weight modulation was characterized using a TCAD device simulation. In addition, we demonstrated the feasibility of the proposed architecture for neuromorphic system applications through a MATLAB simulation for the MNIST pattern recognition. The proposed 3-D synapse array architecture that exhibits a compact chip configuration and a high-integration ability will be a promising technology that can realize hardware-based neuromorphic systems.

Author Contributions: H.-S.C. and Y.K. designed the architecture design and wrote the manuscript. Y.J.P. performed the device simulations. Y.K. confirmed the validities of the designed architecture and simulated synaptic operation. J.-H.L. conceived and developed the various types of 3-D synapse structures, initiated the overall research project. All authors have read and agreed to the published version of the manuscript.

Funding: This work was supported by the 2019 Research Fund of the University of Seoul for Yoon Kim. Also, this work was supported by the MOTIE (Ministry of Trade, Industry & Energy (10080583) and KSRC (Korea Semiconductor Research Consortium) support program for the development of the future semiconductor device for Jong-Ho Lee.

Conflicts of Interest: The authors declare no conflict of interest.

References

1. Yu, S.; Gao, B.; Fang, Z.; Yu, H.; Kang, J.; Wong, H.S. A low energy oxide-based electronic synaptic device for neuromorphic visual systems with tolerance to device variation. *Adv. Mater.* **2013**, *25*, 1774–1779. [CrossRef] [PubMed]
2. Liu, X.; Mao, M.; Liu, B.; Li, H.; Chen, Y.; Li, B.; Wang, Y.; Jiang, H.; Barnell, M.; Wu, Q.; et al. RENO: A high-efficient reconfigurable neuromorphic computing accelerator design. In Proceedings of the 2015 52nd ACM/EDAC/IEEE Design Automation Conference (DAC), San Francisco, CA, USA, 8–12 June 2015; pp. 1–6.
3. Mead, C. Neuromorphic electronic systems. *Proc. IEEE* **1990**, *78*, 1629–1636. [CrossRef]
4. Burr, G.W.; Shelby, R.M.; Sebastian, A.; Kim, S.; Kim, S.; Sidler, S.; Virwani, K.; Ishii, M.; Narayanan, P.; Fumarola, A.; et al. Neuromorphic computing using non-volatile memory. *Adv. Phys. X* **2016**, *2*, 89–124. [CrossRef]
5. Choi, H.S.; Wee, D.H.; Kim, H.; Kim, S.; Ryoo, K.C.; Park, B.G.; Kim, Y. 3-D Floating-Gate Synapse Array with Spike-Time-Dependent Plasticity. *IEEE Trans. Electron. Devices* **2018**, *65*, 101–107. [CrossRef]
6. Roberts, P.D.; Bell, C.C. Spike timing dependent synaptic plasticity in biological systems. *Biol. Cybern.* **2002**, *87*, 392–403. [CrossRef] [PubMed]
7. Akopyan, F.; Sawada, J.; Cassidy, A.; Alvarez-Icaza, R.; Arthur, J.; Merolla, P.; Imam, N.; Nakamura, Y.; Datta, P.; Nam, G.J.; et al. Truenorth: Design and tool flow of a 65 mW 1 million neuron programmable neurosynaptic chip. *IEEE Trans. Comput. Aided Des. Integr. Circuits Syst.* **2015**, *34*, 1537–1557. [CrossRef]
8. Yu, S.; Wu, Y.; Jeyasingh, R.; Kuzum, D.; Wong, H.S. An Electronic Synapse Device Based on Metal Oxide Resistive Switching Memory for Neuromorphic Computation. *IEEE Trans. Electron. Devices* **2011**, *58*, 2729–2737. [CrossRef]
9. Panwar, N.; Kumar, D.; Upadhyay, N.K.; Arya, P.; Ganguly, U.; Rajendran, B. Memristive synaptic plasticity in $Pr_{0.7}Ca_{0.3}MnO_3$ RRAM by bio-mimetic programming. In Proceedings of the 72nd Device Research Conference, Santa Barbara, CA, USA, 22–25 June 2014; pp. 135–136.
10. Diorio, C.; Hasler, P.; Minch, B.A.; Mead, C.A. A floating-gate MOS learning array with locally computed weight updates. *IEEE Trans. Electron. Devices* **1997**, *44*, 2281–2289. [CrossRef]
11. Park, Y.J.; Kwon, H.T.; Kim, B.; Lee, W.J.; Wee, D.H.; Choi, H.S.; Park, B.G.; Lee, J.H.; Kim, Y. 3-D Stacked Synapse Array Based on Charge-Trap Flash Memory for Implementation of Deep Neural Networks. *IEEE Trans. Electron. Devices* **2019**, *66*, 420–427. [CrossRef]
12. Lee, J.; Park, B.G.; Kim, Y. Implementation of Boolean Logic Functions in Charge Trap Flash for In-Memory Computing. *IEEE Electron. Device Lett.* **2019**, *40*, 1358–1361. [CrossRef]
13. Kim, Y.; Kang, M. Down-coupling phenomenon of floating channel in 3D NAND flash memory. *IEEE Electron. Device Lett.* **2016**, *37*, 1566–1569. [CrossRef]
14. Jeong, W.; Im, J.W.; Kim, D.H.; Nam, S.W.; Shim, D.K.; Choi, M.H.; Yoon, H.J.; Kim, D.H.; Kim, Y.S.; Park, H.W.; et al. A 128 Gb 3b/cell V-NAND flash memory with 1 Gb/s I/O rate. *IEEE J. Solid-State Circuits* **2016**, *51*, 204–212.
15. Kang, M.; Kim, Y. Natural Local Self-Boosting Effect in 3D NAND Flash Memory. *IEEE Electron. Device Lett.* **2017**, *38*, 1236–1239. [CrossRef]
16. Canziani, A.; Paszke, A.; Culurciello, E. An Analysis of Deep Neural Network Models for Practical Applications. *arXiv* **2017**, arXiv:1605.07678.
17. *Sentaurus Device User Guide*; Ver. I-2013.12; Synopsys: Mountain View, CA, USA, 2012.
18. Kim, Y.; Seo, J.Y.; Lee, S.H.; Park, B.G. A New Programming Method to Alleviate the Program Speed Variation in Three-Dimensional Stacked Array NAND Flash Memory. *JSTS* **2014**, *5*, 566–571. [CrossRef]

19. LeCun, Y.; Bottou, L.; Bengio, Y.; Haffner, P. Gradient-based learning applied to document recognition. *Proc. IEEE* **1998**, *86*, 2278–2324. [CrossRef]
20. Nair, V.; Hinton, G.E. Rectified linear units improve restricted Boltzmann machines. In Proceedings of the 27th International Conference on Machine Learning (ICML-10), Haifa, Israel, 21–24 June 2010; pp. 807–814.
21. Burr, G.W.; Shelby, R.M.; Sidler, S.; Di Nolfo, C.; Jang, J.; Boybat, I.; Shenoy, R.S.; Narayanan, P.; Virwani, K.; Giacometti, E.U.; et al. Experimental Demonstration and Tolerancing of a Large-Scale Neural Network (165000 Synapses) Using Phase-Change Memory as the Synaptic Weight Element. *IEEE Trans. Electron. Devices* **2015**, *62*, 3498–3507. [CrossRef]
22. Kim, H.; Hwang, S.; Park, J.; Park, B.G. Silicon synaptic transistor for hardware-based spiking neural network and neuromorphic system. *Nanotechnology* **2017**, *28*, 405202. [CrossRef] [PubMed]
23. Kim, S.; Kim, H.; Hwang, S.; Kim, M.H.; Chang, Y.F.; Park, B.G. Analog Synaptic Behavior of a Silicon Nitride Memristor. *ACS Appl. Mater. Interfaces* **2017**, *9*, 40420–40427. [CrossRef] [PubMed]
24. Ambrogio, S.; Ciocchini, N.; Laudato, M.; Milo, V.; Pirovano, A.; Fantini, P.; Ielmini, D. Unsupervised learning by spike timing dependent plasticity in phase change memory (PCM) synapses. *Front. Neurosci.* **2016**, *10*, 56. [CrossRef] [PubMed]

Article

Control of the Boundary between the Gradual and Abrupt Modulation of Resistance in the Schottky Barrier Tunneling-Modulated Amorphous Indium-Gallium-Zinc-Oxide Memristors for Neuromorphic Computing

Jun Tae Jang †, Geumho Ahn †, Sung-Jin Choi, Dong Myong Kim and Dae Hwan Kim *

School of Electrical Engineering, Kookmin University, Seoul 02707, Korea; jtjang@kookmin.ac.kr (J.T.J.); 20113145@kookmin.ac.kr (G.A.); sjchoiee@kookmin.ac.kr (S.-J.C.); dmkim@kookmin.ac.kr (D.M.K.)

* Correspondence: drlife@kookmin.ac.kr; Tel.: +82-2-910-4872

† These authors are co-first author.

Received: 4 September 2019; Accepted: 23 September 2019; Published: 25 September 2019

Abstract: The transport and synaptic characteristics of the two-terminal Au/Ti/ amorphous Indium-Gallium-Zinc-Oxide (a-IGZO)/thin SiO_2/p^+-Si memristors based on the modulation of the Schottky barrier (SB) between the resistive switching (RS) oxide layer and the metal electrodes are investigated by modulating the oxygen content in the a-IGZO film with the emphasis on the mechanism that determines the boundary of the abrupt/gradual RS. It is found that a bimodal distribution of the effective SB height (Φ_B) results from further reducing the top electrode voltage (V_{TE})-dependent Fermi-level (E_F) followed by the generation of ionized oxygen vacancies ($V_O{}^{2+}$s). Based on the proposed model, the influences of the readout voltage, the oxygen content, the number of consecutive V_{TE} sweeps on Φ_B, and the memristor current are explained. In particular, the process of $V_O{}^{2+}$ generation followed by the Φ_B lowering is gradual because increasing the V_{TE}-dependent E_F lowering followed by the $V_O{}^{2+}$ generation is self-limited by increasing the electron concentration-dependent E_F heightening. Furthermore, we propose three operation regimes: the readout, the potentiation in gradual RS, and the abrupt RS. Our results prove that the Au/Ti/a-IGZO/SiO_2/p^+-Si memristors are promising for the monolithic integration of neuromorphic computing systems because the boundary between the gradual and abrupt RS can be controlled by modulating the SiO_2 thickness and IGZO work function.

Keywords: a-IGZO memristor; Schottky barrier tunneling; non filamentary resistive switching; gradual and abrupt modulation; bimodal distribution of effective Schottky barrier height; ionized oxygen vacancy

1. Introduction

The electronic computing systems developed so far have been structured on the von Neumann architecture in which the memory, the processor, and the controller exist separately, and the sequential processing among them embodies specific functions within the programmed software. Most of the digital and analog circuits included in the memory and processing units are composed of complementary metal-oxide-semiconductor (CMOS) devices that have made a significant contribution to the semiconductor industry. Improvements in the performance of modern computing and information technology are based on the permanent scaling down of the CMOS devices, which provide a cost-effective increase in the operating frequency and a reduction in the power consumption [1,2].

Currently, the integration density of CMOS devices do not conform to Moore's law [3], and the scaling down is fast approaching the physical limit. However, an increase in the operating frequency and the device density increases the power consumption and the operation temperature, which can seriously degrade the system performance (von Neumann bottleneck), mainly because of the time and energy spent in transporting data between the memory and the processor [4]. This is particularly noticeable for data-centric applications, such as real-time image recognition and natural language processing, where the state-of-the-art von Neumann systems cannot outperform an average human.

Unlike with the von Neumann systems, the human brain creates a massively parallel architecture by connecting a large number of low-power computing elements (neurons) and adaptive memory elements (synapses). Thus, the brain can outperform modern processors on many tasks that involve unstructured data classification and pattern recognition [5]. Furthermore, the ultra-dense crossbar array consisting of memristors have been recognized as a potentially promising path to building neuromorphic computing systems that can mimic the massive parallelism and extremely low-power operations found in the human brain [6]. Representative types of neuromorphic computing schemes are the biologically inspired spiking neural networks (SNNs) and deep neural networks, which are vector matrix multipliers [7,8]. The SNNs are based on the local spike-timing-dependent plasticity (STDP) learning rule [7], whereas the latter is based on the backpropagation learning rule [8].

The two-terminal binary metal-oxide-based resistive switching (RS) devices, such as HfO_x, AlO_x, WO_x, TaO_x, and TiO_x, have been widely studied as memristor devices that play the role of synapses in the crossbar arrays because the underlying metal–insulator–metal structure is simple, compact, CMOS-compatible, and highly scalable. Indeed, their energy consumptions per synaptic operation and programming currents can be made ultralow (sub-pJ energies, <1 μA programming current) [9]. However, in most cases of these *filamentary* resistive switching random access memory (hereinafter ReRAM) devices, the filament formation/completion process is inherently abrupt and difficult to control. This problem is particularly acute in neuromorphic applications because a single highly conductive device with a thick filament provides much more current to a vector-weighted sum or a leaky integrate-and-fire than its neighbors [10]. Undoubtedly, the gradual RS characteristics (i.e., the analog nonvolatile memory characteristics of the memristors) are most viable for either the weighted sum operation of convolutional neural networks (CNNs) or the STDP as a learning rule for SNN. In particular, the synapse device using the memristor requires excellent linearity according to the consecutive potentiation/depression pulse for high data processing accuracy [11].

In the case of filamentary ReRAM devices, there is ambiguity at the boundary between the application of the digital memory device using the abrupt RS operation and the application of the synapse device using the gradual RS operation. Therefore, it is very difficult to optimize each of the devices for both applications in terms of the process and the material. More noticeably, the efficiency and linearity of the resistance modulations of the metal-oxide-based memristors are frequently contradictory to one another when applying the potentiation/depression (P/D) pulses [12]. This is because when the resistance changes of the filamentary ReRAM devices occur more efficiently (abruptly), the resistances become more nonlinear in relation to the increase in the number of P/D pulses. After being triggered by an electric field and/or a local temperature rise during the SET/potentiation pulse, the filament formation/completion must be cut by an external circuit so that the filament is not too thick to be removed with an accessible RESET/depression pulse. Despite using techniques such as incrementally increasing the amplitude of the P/D voltage and/or increasing the duration of the P/D pulse [13], the complicated scheme for self-adaptively varying either the amplitude or the duration of the P/D pulse would be significantly compromised with the use of external controls and circuits. This results in additional power consumption and design complexity and seriously dilutes the motivation of neuromorphic computing systems.

However, *non-filamentary* RS two-terminal devices based on binary metal-oxides have demonstrated more gradual (well-controlled, memristor-like) RS characteristics in comparison with filamentary RS devices [14] because the non-filamentary devices are based on the modulation of the Schottky barrier (SB) between the RS oxide layer and the metal electrodes rather than the formation/rupture of the filament in the oxide layer.

Regardless of the type of RS devices, for a systematic and robust design of a self-adaptive P/D pulse scheme, it is important to have a complete understanding of the physical mechanism that controls the boundary of an abrupt/gradual RS characteristic. Therefore, it is important to understand the systematic design of the memristor devices for neuromorphic computing and precisely control the mechanism on the boundary of the abrupt and the gradual RS operations.

Quaternary metal-oxides, such as amorphous indium-gallium-zinc-oxide (a-IGZO), have more complicated compositions and they cannot be easily fabricated by low-temperature sputtering or the solution process. The a-IGZO materials can be fabricated on a flexible substrate and can act as both the RS and active films in memristors and thin-film transistors (TFTs), respectively [15–20]; this suggests that it is possible to monolithically integrate not only the synapse array but also the peripheral circuits including the neurons. In fact, two-terminal IGZO devices and their abrupt/gradual switching characteristics using metal electrodes, such as Pt, Al, and Cu, have already been demonstrated [16–20]. Even unipolar/bipolar IGZO memristor devices have been developed [19,20]. However, there is no known mechanism for determining the boundary of an abrupt/gradual RS in IGZO memristor devices.

In this study, we fabricated two-terminal Au/Ti/a-IGZO/thin SiO_2/p^+-Si memristors and analyzed their transport and synaptic characteristics. Moreover, we investigated the mechanism determining the boundary of the abrupt/gradual RS by modulating the oxygen content in an a-IGZO film. Related to this mechanism, we also reported a bimodal distribution of effective Schottky barriers in a-IGZO non-filamentary ReRAM-based memristors.

2. Fabrication Process and Conduction Mechanism

To implement the synapse devices in bio-inspired neuromorphic computing systems (Figure 1a), we fabricated the two-terminal Au/Ti/IGZO/SiO_2/p^+-Si memristors as shown Figure 1b. The p^+-Si conductive substrate acts as a global bottom electrode (BE), and the 4-nm-thick SiO_2 was formed on the BE as the tunnel barrier in the interface between p^+-Si and IGZO. Then, the 80-nm-thick a-IGZO film was deposited on SiO_2/p^+-Si using radio frequency sputtering with a power of 150 W at room temperature. We controlled the concentration of oxygen vacancies (V_Os) during the IGZO sputtering by modulating the oxygen flow rates (OFR) to 1.0, 1.15, and 1.3 sccm at a fixed Ar flow rate of 3 sccm and at a constant gas pressure in the sputter chamber of 0.880 Pa. Subsequently, 10-nm-thick Ti was deposited using e-beam evaporation to form an oxygen reservoir layer and act as the top electrode (TE) of the memristor. Finally, the 40-nm-thick Au was deposited using e-beam evaporation to prevent the oxidation of the Ti layer in air.

To analyze the electrical characteristics, the DC current–voltage ($I-V$) characteristics were measured at room temperature and dark conditions using a Keithley-4200 semiconductor characterization system (Tektronix, Seoul, South Korea). In all the measurements, a voltage was applied to the TE, and the BE was always connected to the ground. The TE voltage was symbolized as V_{TE}, and the current flowing through the IGZO memristor was called I_{mem}, as shown in Figure 1b.

Figure 1c–f shows the energy band diagrams under various conditions: before forming the junction (Figure 1c), at the thermal equilibrium (Figure 1d), at a low V_{TE} (Figure 1e), and at a high V_{TE} (Figure 1f). Here, we considered the lowering of the height of the effective SB and denoted it as $q\Phi_B$ (eV). While SB lowering was insignificant at a thermal equilibrium, $q\Phi_B$ became low as the V_{TE} increased. At a low V_{TE}, most of the V_{TE} was applied across the thin SiO_2 layer (Figure 1e), whereas the increased V_{TE} was used mainly to deplete the IGZO film (Figure 1f). Energy band diagrams suggested the fabricated IGZO memristors operated as non-filamentary RS devices based on the SB modulation. The two main concerns were whether the modulated $q\Phi_B$ was nonvolatile and whether its decrease was inversely linear with the increase of V_{TE}. These two concerns will be discussed later.

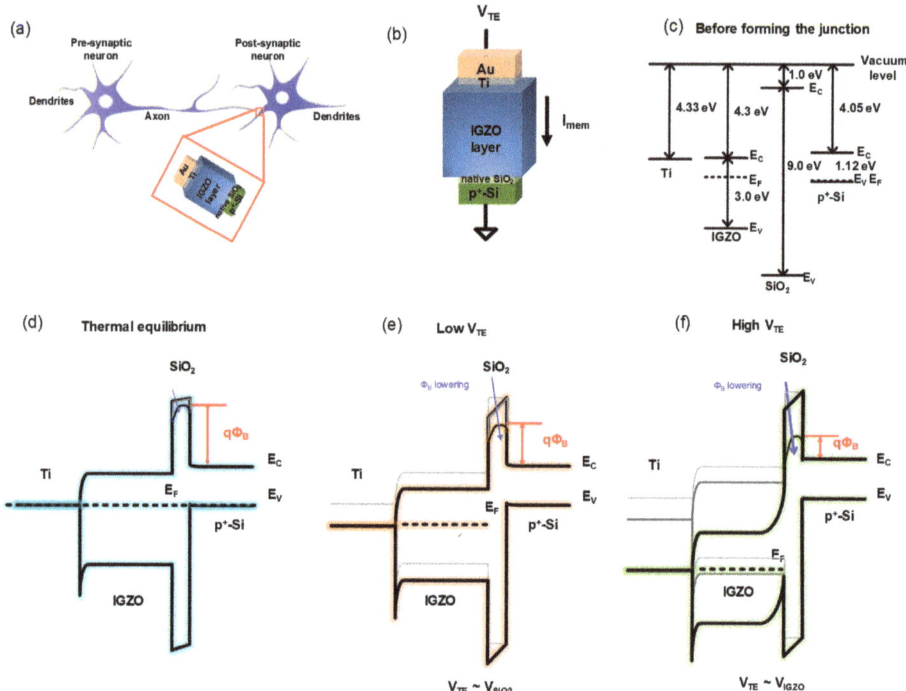

Figure 1. Schematic illustration of (**a**) the implementation of the synapse devices in bio-inspired neuromorphic computing systems and (**b**) the two-terminal Au/Ti/(amorphous indium-gallium-zinc-oxide) a-IGZO/SiO_2/p^+-Si memristors. Energy band diagram (**c**) before forming the junction, and under three conditions: (**d**) in a thermal equilibrium, (**e**) at a low (top electrode voltage) V_{TE}, and (**f**) at a high V_{TE}.

We measured the OFR-dependent I_{mem} while using a positive V_{TE} sweep (SET process), that is, 0 V → 6 V → 0 V was repeated four times. Then, a negative V_{TE} sweep (RESET process), that is, 0 V → −2 V → 0 V was repeated four times, as shown in Figure 2a. We observed that the current at a fixed V_{TE} increased as the OFR decreased. This was attributed to the increase of the V_O concentration with the decrease in the OFR because the V_O is a well-known electron donor in the IGZO film [21,22]. Along with the SB-modulated non-filamentary RS devices in Figure 1e,f, a gradual resistance modulation rather than an abrupt RS was clearly observed during repeated I–V sweeps (Figure 2a).

Figure 2b also shows the I_{mem}–V_{TE} characteristic of the IGZO memristor with OFR = 1 sccm. In Figure 2b, the positive V_{TE} voltage sweep was repeated four consecutive times by changing the stop voltage of the V_{TE} sweep (V_{SS}) from 2 to 6 V. When the V_{TE} sweep was performed four times, the readout current I_{mem} at V_{TE} = 1 V increased very slightly for $V_{SS} < 6$ V, as seen in Figure 2c. The continuous and hysteretic increase of current, which is a typical behavior of a memristor, is clearly observed in Figure 2a,b. There was a significant increase in I_{mem} only when $V_{SS} \geq 6$ V, which means that the *potentiation threshold* voltage between the gradual/abrupt RS (V_{PT}) was 6 V. Similarly, the *depression threshold* voltage was found to be −2 V.

To determine the conduction mechanism, we investigated the relationship between I_{mem} and V_{TE}. Figure 3a shows the OFR-dependent ln(I_{mem}) versus $(V_{TE})^{1/2}$ relationships, which were taken from the I–V characteristics of the first sweep in Figure 2a. In Figure 3a, we observed that the ln(I_{mem}) was piecewise linear with $(V_{TE})^{1/2}$, which was strongly reminiscent of the thermionic emission. Noticeably, these linear relationships were clearly classified into two distinguishable values of the slopes A (at a low V_{TE}) and B (at a high V_{TE}).

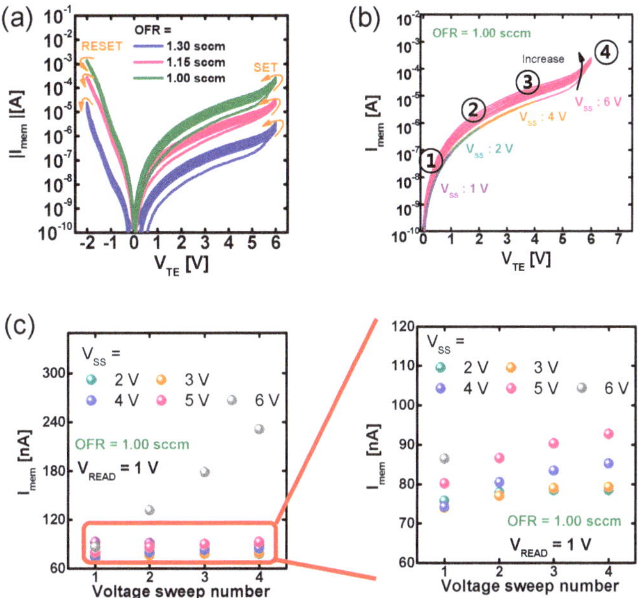

Figure 2. (**a**) The (oxygen flow rate) OFR-dependent I–V characteristics repeated four times. (**b**) The I–V characteristics with OFR = 1 sccm repeated four consecutive times with changes made to the (stop voltage of the V_{TE} sweep) V_{SS}. (**c**) The V_{SS}-dependent readout current I_{mem} at V_{TE} = 1 V.

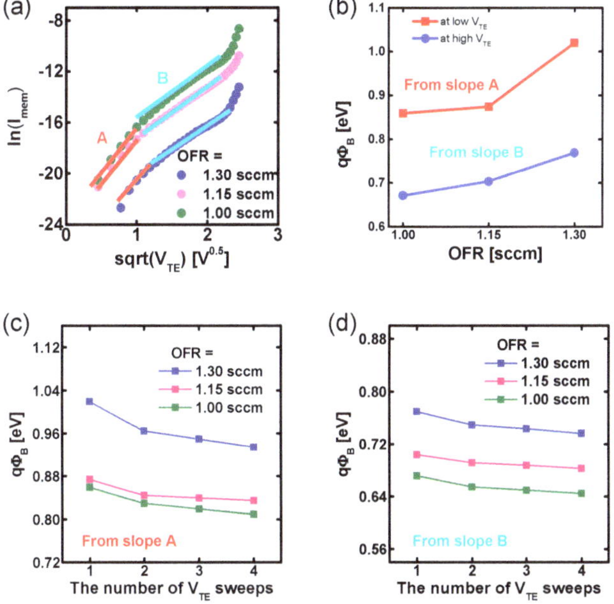

Figure 3. (**a**) The OFR-dependent $\ln(I_{mem})$–$(V_{TE})^{1/2}$ relationships. (**b**) The OFR-dependent Φ_B extracted at low and high V_{TE}. The Φ_B modulation depending on the number of V_{SS} from the slopes (**c**) A at high V_{TE} and (**d**) B at low V_{TE}.

The current due to the thermionic emission through SB is given as:

$$I_{\text{mem}} = AA^{*}T^{2}\exp\left(\frac{q(\sqrt{qE/4\pi\varepsilon} - \Phi_{\text{B}})}{kT}\right) = AA^{*}T^{2}\exp\left(\frac{q(\sqrt{qV_{TE}/4\pi\varepsilon X_{T}} - \Phi_{\text{B}})}{kT}\right) \tag{1}$$

where A is the area of device, A^* is the Richardson constant, T is the absolute temperature, k is Boltzmann's constant, E is the electric field; q is the electric charge, ε is the dielectric constant, X_{T} is the effective thickness of thermionic emission, and Φ_{B} is the effective SB height. Then, Equation (1) is used for extracting Φ_{B}. By reformulating from Equation (1) to (2), Φ_{B} can be extracted by using the y-intercept of the linear relationship between $\frac{kT}{q}\cdot\ln\left(\frac{I_{\text{mem}}}{AA^{*}T^{2}}\right)$ and $\sqrt{V_{TE}}$:

$$\frac{kT}{q}\cdot\ln\left(\frac{I_{\text{mem}}}{AA^{*}T^{2}}\right) = \sqrt{\frac{q/X_{T}}{4\pi\varepsilon}} \times \sqrt{V_{TE}} - \Phi_{\text{B}} \tag{2}$$

Figure 3a,b suggests that at a specific OFR, there existed two Φ_{B} values taken from the slopes A and B, that is, a large value for a low V_{TE} (<1 V) and a small value for a high V_{TE} (1–5 V). Interestingly, we observed this bimodal distribution of Φ_{B} regardless of the OFR condition and suggest that the SB lowering is nonvolatile and significantly nonlinear with the increase in V_{TE}. In addition, Φ_{B} at a specific V_{TE} was lower because the V_{O} concentration increases (with decreasing OFR).

However, from Figure 2a, we can see that the Φ_{B} modulation depended on the number of positive V_{TE} sweeps (see Figure 3c,d). At a specific V_{TE} and OFR, Φ_{B} gradually decreased when the number of V_{SS} sweeps increased.

3. Results and Discussion

In Figure 3, we can see that Φ_{B} was modulated by not only the range of the V_{TE} readout voltage, but also by the number of consecutive V_{SS} sweeps. Moreover, as shown in Figure 3b,c, Φ_{B} depends more strongly on OFR in the slope A case (low V_{TE}) rather than in the slope B case (high V_{TE}). Therefore, the results in Figure 3 provide a clue toward the controllability of the competition between the gradual and abrupt modulations of Φ_{B}. To understand the mechanism for determining the boundary of an abrupt/gradual RS in IGZO memristor devices, we used Figure 3 with the energy band diagram.

First, when $V_{\text{TE}} < V_{\text{PT}}$, the bimodal distribution of Φ_{B} into A and B (Figure 3a) can be explained as follows. As shown in Figure 4a, the doubly ionized V_{O} (V_{O}^{2+}) is the well-known metastable state [21,22] and has been frequently pointed out as having a microscopic origin on the device instability under photo-illumination or bias stress [22–26] and persistent photoconductivity [25,26]. From the viewpoint of the subgap density of states (DOSs) in the a-IGZO (Figure 4b), the neutral V_{O} states (V_{O}^{0}s) are transformed into V_{O} (V_{O}^{2+}s) when the process of $V_{\text{O}}^{0} \rightarrow V_{\text{O}}^{2+} + 2e^{-}$ becomes energetically favorable. These neutral states are very slowly recovered (nonvolatile) [23–26].

In the readout voltage V_{TE}-dependent energy band diagrams, which are illustrated in Figure 4c, as V_{TE} increases, the Fermi-energy level (E_{F}) in IGZO reduces far from the IGZO conduction band minimum (E_{C}), and moves closer to the V_{O}^{0} states above the IGZO valence band maximum (E_{V}). It makes the generation of V_{O}^{2+}s more energetically favorable. When V_{O}^{2+}s is generated, the concentration of the carrier electrons in E_{C} increases; the E_{F} in IGZO again comes closer to E_{C}. This situation occurs in non-equilibrium; therefore, the generation of V_{O}^{2+}s effectively makes Φ_{B} lower.

Thus, if the V_{O} ionization is nonvolatile, Φ_{B} would gradually decrease as the readout voltage V_{TE} increases. In other words, Φ_{B} has to be inversely linear to V_{TE}. However, Φ_{B} was classified into two groups (A and B), as seen in Figure 3. Figure 1e,f shows that a large Φ_{B} (in low V_{TE}) taken from the slope A corresponded to the voltage range where the maximum V_{TE} was applied across a thin SiO_2 layer (Figure 1e), whereas a small Φ_{B} (in high V_{TE}) taken from the slope B corresponded to the voltage range where the maximum increase in V_{TE} was mainly applied across the IGZO film (Figure 1f). Then, there would be a significant generation of V_{O}^{2+}s only in the latter range (Figure 4c). In Figure 2c, I_{mem} gradually increased only when V_{TE} was in the latter range, that is, in the range 2 V $\leq V_{\text{TE}} < V_{\text{PT}}$.

Our discussion indicates that the bimodal distribution of Φ_B in IGZO memristors originated from the generation of metastable $V_O{}^{2+}$ states.

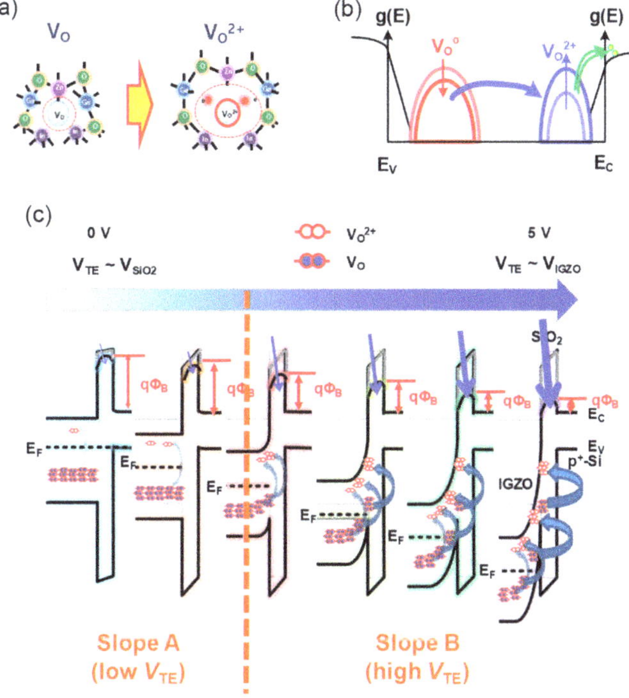

Figure 4. Schematic illustration of oxygen vacancies ionization from the viewpoint of (**a**) the atomic structures and (**b**) the subgap DOS in a-IGZO. (**c**) The read voltage V_{TE}-dependent energy band diagram.

Next, we investigated the OFR-dependence of Φ_B. Figure 5a–c illustrates the energy band diagram of the device fabricated with a high OFR (O-rich device) under three conditions: at a thermal equilibrium (Figure 5a), at a low V_{TE} (Figure 5b), and at a high V_{TE} (Figure 5c). Figure 5d–f illustrates the energy band diagram of the device fabricated using a low OFR (O-poor device) in three states: at a thermal equilibrium (Figure 5d), at a low V_{TE} (Figure 5e), and at a high V_{TE} (Figure 5f). As seen in Figure 5a,d, a larger amount of $V_O{}^0$s existed in the IGZO when the OFR decreased from 1.3 to 1.0 sccm. Then, as the IGZO was O-poorer, the IGZO work function decreased, and Φ_B became lower, which is consistent with Figure 3b. In addition, as mentioned in Figure 3b,c, the OFR-dependence of Φ_B was larger in the slope A case (low V_{TE}) rather than in the slope B case (high V_{TE}). The Φ_B before the $V_O{}^{2+}$ generation (at a low V_{TE}) was determined mainly by the OFR condition. After a significant amount of $V_O{}^{2+}$s were generated at a high V_{TE}, the initial OFR-dependence of Φ_B was combined with the V_{TE}-dependence of Φ_B. Thus, the OFR-dependence of Φ_B was diluted in the slope B case (high V_{TE}).

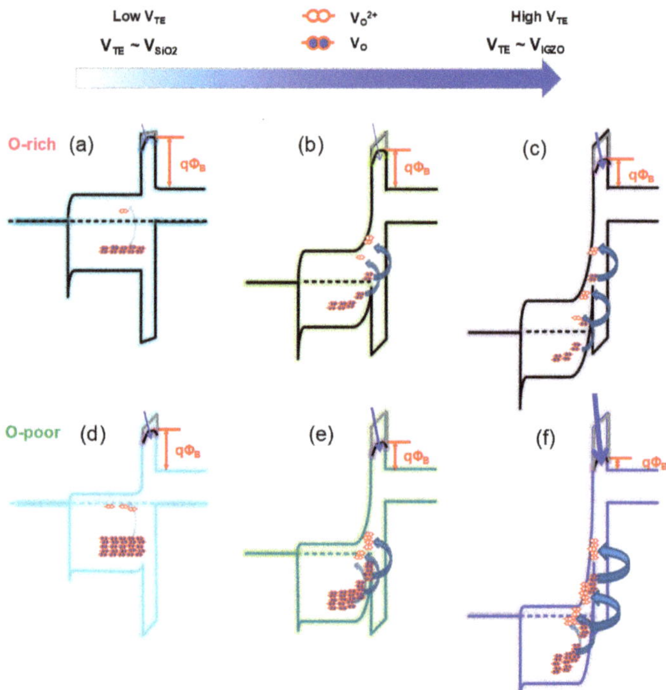

Figure 5. The OFR-dependent energy band diagram and Φ_B with (**a**)–(**c**) high OFR and (**d**)–(**f**) low OFR at (**a**,**d**) thermal equilibrium, (**b**,**e**) low V_{TE}, and (**c**,**f**) high V_{TE}.

Finally, the evolution of Φ_B with the increase in the number of consecutive positive V_{SS} sweeps is illustrated in the energy band diagrams in Figure 6. When the V_{SS} sweeps were repeated four times, Φ_B gradually decreased because of the gradual increase in V_O^{2+}s. However, the process of V_O^{2+} generation followed by Φ_B lowering was not abrupt; it was gradual because further lowering of the V_{TE}-dependent E_F followed by the V_O^{2+} generation was self-limited due to the increasing of the electron concentration–dependent E_F. The results in Figure 3c,d explain this well. If $V_{TE} \geq V_{PT}$, the change of I_{mem} becomes abrupt because E_F is aligned with the level of the V_O^0s peak in DOS (Figure 4b).

Therefore, we can classify the operation regime in the two-terminal Au/Ti/a-IGZO/SiO_2/p^+-Si memristors into three parts: (1) low V_{TE} ($V_{TE} < 2$ V), (2) high V_{TE} (2 V $\leq V_{TE} \leq V_{PT}$), and (3) higher V_{TE} ($V_{TE} \geq V_{PT}$). The boundary between (1) and (2) was approximately 2 V in our case; it was determined by the process/structure details and was controllable using the SiO_2 thickness and the IGZO work function. The V_{TE} in regime (1) was adequate for the readout voltage because Φ_B and I_{mem} were determined mainly by the OFR condition. However, the V_{TE} in regime (2) can be used as the amplitude of the potential pulse because Φ_B and I_{mem} gradually change in a nonvolatile manner with the increase in the number of consecutive V_{SS} sweeps. When the V_{TE} in regime (3) was applied to the devices, they operated as abrupt RS switches rather than as gradual RS memristors.

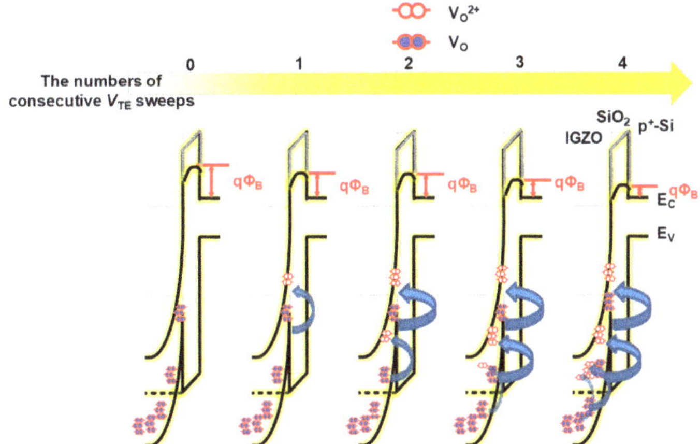

Figure 6. Energy band diagram for the evolution of Φ_B with the increase in the number of consecutive positive V_{SS} sweeps.

4. Conclusions

It is crucial to have good control over the mechanism on the boundary between the abrupt and gradual RS operations for a systematic design of memristor devices for neuromorphic computing. We investigated the transport and synaptic characteristics of two-terminal Au/Ti/a-IGZO/thin SiO_2/p$^+$-Si memristors by varying the oxygen content in the a-IGZO film by emphasizing the mechanism determining the boundary of the abrupt/gradual RS. A bimodal distribution of Φ_B was produced to further lower the V_{TE}-dependent E_F followed by the generation of V_O^{2+}s. Based on the proposed model, we explained the influence of the readout voltage, the oxygen content, and the number of consecutive V_{SS} sweeps on Φ_B and I_{mem}. Eventually, we proposed three operation regimes: the readout, the potentiation in gradual RS, and the abrupt RS.

Our results prove that the Au/Ti/a-IGZO/SiO_2/p$^+$-Si memristors are promising for the monolithic integration of neuromorphic computing systems because the boundary between the gradual and the abrupt RS can be controlled by modulating the SiO_2 thickness and the IGZO work function. Furthermore, the memristors are expected to be potentially useful for the co-design and joint optimization of the IGZO memristors and TFTs for neuromorphic energy-efficient wearable healthcare circuits and systems.

Author Contributions: The manuscript was prepared by J.T.J., G.A., S.-J.C., D.M.K., and D.H.K. Device fabrication was performed by J.T.J. and G.A. Results and discussion were performed by J.T.J., G.A., and D.H.K.

Funding: This work was supported by the national research foundation (NRF) of Korea funded by the Korean government under Grant 2016R1A5A1012966, 2016M3A7B4909668, 2017R1A2B4006982, and in part by an Electronics and Telecommunications Research Institute (ETRI) grant funded by the Korean government (18ZB1800).

Conflicts of Interest: The authors declare no conflict of interest.

References

1. Wan, Q.; Sharbati, M.T.; Erickson, J.R.; Du, Y.; Xiong, F. Emerging Artificial Synaptic Device for Neuromorphic Computing. *Adv. Mater. Technol.* **2019**, *4*, 1900037. [CrossRef]
2. Upadhyay, N.K.; Jiang, H.; Wang, Z.; Asapu, S.; Xia, Q.; Yang, J.J. Emerging Memory Devices for Neuromorphic Computing. *Adv. Mater. Technol.* **2019**, *4*, 1800589. [CrossRef]
3. Kish, L.B. End of Moore's law: Thermal (noise) death of integration in mirco and nano electronics. *Phys. Lett. A* **2002**, *305*, 144–149. [CrossRef]
4. Neumann, J.V. First Draft of a Report on the EDVAC. *Ann. Hist. Comput.* **1993**, *15*, 27–75. [CrossRef]

5. Indiveri, G.; Liu, S.-C. Memory and Information Processing in Neuromorphic Systems. *Proc. IEEE.* **2015**, *103*, 1379–1397. [CrossRef]
6. Park, S.; Kim, H.; Choo, M.; Noh, J.; Sheri, A.; Jung, S.; Seo, K.; Park, J.; Kim, S.; Lee, W.; et al. RRAM-based Synapse for Neuromorphic System with Pattern Recognition Function. In Proceedings of the 2012 International Electron Devices Meeting, San Francisco, CA, USA, 10−13 December 2012; pp. 10.2.1–10.2.4.
7. Kheradpisheh, S.R.; Ganjtabesh, M.; Thorpe, S.J.; Masquelier, T. STDP-based spiking deep convolutional neural networks for object recognition. *Neural Netw.* **2018**, *99*, 56–67. [CrossRef] [PubMed]
8. LeCun, Y.; Bengio, Y.; Hinton, G. Deep learning. *Nature* **2015**, *521*, 436–444. [CrossRef] [PubMed]
9. Yu, S.; Gao, B.; Fang, Z.; Yu, H.; Kang, J.; Wong, H.-S.P. A Neuromorphic Visual System Using RRAM Synaptic Devices with Sub-pJ Energy and Tolerance to Variability: Experimental Characterization and Large-Scale Modeling. In Proceedings of the 2012 International Electron Devices Meeting, San Francisco, CA, USA, 10−13 December 2012; pp. 10.4.1–10.4.4.
10. Burr, G.W.; Shelby, R.M.; Sebastian, A.; Kim, S.; Kim, S.; Sidler, S.; Virwani, K.; Ishii, M.; Narayanan, P.; Fumarola, A. Neuromorphic computing using non-volatile memory. *Adv. Phys. X* **2016**, *2*, 89–124. [CrossRef]
11. Yu, S. Neuro-Inspired Computing with Emerging Nonvolatile Memory. *Proc. IEEE* **2018**, *106*, 260–285. [CrossRef]
12. Kim, S.; Choi, B.; Lim, M.; Yoon, J.; Lee, J.; Kim, H.-D.; Choi, S.-J. Pattern Recognition Using Carbon Nanotube Synaptic Transistors with an Adjustable Weight Update Protocol. *ACS Nano.* **2017**, *11*, 2814–2822. [CrossRef]
13. Jerry, M.; Chen, P.-Y.; Zhang, J.; Sharma, P.; Ni, K.; Yu, S.; Datta, S. Ferroelectric FET Analog Synapse for Acceleration of Deep Neural Network Training. In Proceedings of the 2017 International Electron Devices Meeting, San Francisco, CA, USA, 2–6 December 2017; pp. 6.2.1–6.2.4.
14. Woo, J.; Moon, K.; Song, J.; Lee, S.; Kwak, M.; Park, J.; Hwang, H. Improved Synaptic Behavior Under Identical Pulses Using AlO_x/HfO_2 Bilayer RRAM Array for Neuromorphic Systems. *IEEE Electron Device Lett.* **2016**, *37*, 994–997. [CrossRef]
15. Boldman, W.L.; Zhang, C.; Ward, T.Z.; Briggs, D.P.; Srijanto, B.R.; Brisk, P.; Rack, P.D. Programmable Electrofluidics for Ionic Liquid Based Neuromorphic Platform. *Micromachines* **2019**, *10*, 478. [CrossRef] [PubMed]
16. Dang, B.; Liu, K.; Zhu, J.; Xu, L.; Zhang, T.; Cheng, C.; Wang, H.; Yang, Y.; Hao, Y.; Huang, R. Stochastic neuron based on IGZO Schottky diodes for neuromorphic computing. *APL Mater.* **2019**, *7*, 071114. [CrossRef]
17. Wang, Z.Q.; Xu, H.Y.; Li, X.H.; Yu, H.; Liu, Y.C.; Zhu, X.J. Synaptic Learning and Memory Functions Achieved Using Oxygen Ion Migration/Diffusion in an Amorphous InGaZnO Memristor. *Adv. Funct. Mater.* **2012**, *22*, 2759–2765. [CrossRef]
18. Kado, K.; Uenuma, M.; Sharma, K.; Yamazaki, H.; Urakawa, S.; Ishikawa, Y.; Uraoka, Y. Thermal analysis for observing conductive filaments in amorphous InGaZnO thin film resistive switching memory. *Appl. Phys. Lett.* **2014**, *105*, 123506. [CrossRef]
19. Hu, W.; Zou, L.; Chen, X.; Qin, N.; Li, S.; Bao, D. Highly Uniform Resistive Switching Properties of Amorphous InGaZnO Thin Films Prepared by a Low Temperature Photochemical Solution Deposition Method. *ACS Appl. Mater. Interfaces* **2014**, *6*, 5012–5017. [CrossRef]
20. Wang, Z.; Xu, H.; Zhao, X.; Lin, Y.; Zhang, L.; Ma, J.; Liu, Y. Effect of reset voltage polarity on the resistive switching region of unipolar memory. *Phys. Status Solidi A* **2015**, *212*, 2255–2261. [CrossRef]
21. Janotti, A.; Van de Walle, C.G. Native point defects in ZnO. *Phys. Rev. B* **2007**, *76*, 165202. [CrossRef]
22. Migliorato, P.; Chowdhury, M.D.H.; Um, J.G.; Seok, M.; Jang, J. Light/negative bias stress instabilities in indium gallium zinc oxide thin film transistors explained by creation of a double donor. *Appl. Phys. Lett.* **2010**, *97*, 022108. [CrossRef]
23. Jang, J.T.; Park, J.; Ahn, B.D.; Kim, D.M.; Choi, S.-J.; Kim, H.-S.; Kim, D.H. Effect of direct current sputtering power on the behavior of amorphous indium-gallium-zinc-oxide thin-film transistors under negative bias illumination stress: A combination of experimental analyses and device simulation. *Appl. Phys. Lett.* **2015**, *106*, 123505. [CrossRef]
24. Hoshino, K.; Wager, J. Negative bias illumination stress assessment of indium gallium zinc oxide thin-film transistors. *J. Soc. Inf. Disp.* **2015**, *23*, 187–195. [CrossRef]

25. Jang, J.T.; Park, J.; Ahn, B.D.; Kim, D.M.; Choi, S.-J.; Kim, H.-S.; Kim, D.H. Study on the Photoresponse of Amorphous In–Ga–Zn–O and Zinc Oxynitride Semiconductor Devices by the Extraction of Sub-Gap-State Distribution and Device Simulation. *ACS Appl. Mater. Interfaces* **2015**, *7*, 15570–15577. [CrossRef] [PubMed]
26. Jeon, S.; Ahn, S.-E.; Song, I.; Kim, C.J.; Chung, U.-I.; Lee, E.; Yoo, I.; Nathan, A.; Lee, S.; Robertson, J.; et al. Gated three-terminal device architecture to eliminate persistent photoconductivity in oxide semiconductor photosensor arrays. *Nat. Mater.* **2012**, *11*, 301–305. [CrossRef] [PubMed]

Article

Analysis of the Voltage-Dependent Plasticity in Organic Neuromorphic Devices

Seunghyuk Lee and Chang-Hyun Kim *

Department of Electronic Engineering, Gachon University, Seongnam 13120, Korea; dobby0227@gc.gachon.ac.kr
* Correspondence: chang-hyun.kim@gachon.ac.kr

Received: 15 November 2019; Accepted: 17 December 2019; Published: 18 December 2019

Abstract: The bias-dependent signal transmission of flexible synaptic transistors is investigated. The novel neuromorphic devices are fabricated on a thin and transparent plastic sheet, incorporating a high-performance organic semiconductor, dinaphtho[2,3-b:2′,3′-f]thieno[3,2-b]thiophene, into the active channel. Upon spike emulation at different synaptic voltages, the short-term plasticity feature of the devices is substantially modulated. By adopting an iterative model for the synaptic output currents, key physical parameters associated with the charge carrier dynamics are estimated. The correlative extraction approach is found to yield the close fits to the experimental results, and the systematic evolution of the timing constants is rationalized.

Keywords: flexible electronics; neuromorphic engineering; organic field-effect transistors; synaptic devices; short-term plasticity

---

1. Introduction

Neuromorphic engineering is an emerging technological area, which aims at mimicking the biological functionalities of neurons, synapses, or a whole brain by various electronic materials and devices [1–6]. Recently, the use of organic electronics in neuromorphic systems has gained tremendous attention, thanks to its capacity to expand the technological scope of such systems by creating unconventional interfaces such as direct neuroprotheses and robotic sensory bridges [7–10]. There are many possible routes to organic-based neuromorphic architecture, including electrochemical [11,12], memristive [13], and field-effect approaches [14–16]. Among them, organic field-effect transistor (OFET)-based synaptic devices are a particularly promising element, considering the possibility of a fully solid-state, flexible neuromorphic chip that leverages the versatility of OFETs in constructing various circuit building blocks [17–20]. Despite the rapidly growing technological viability of OFET synapses, there is still a lack of understanding on fundamental phenomena prevailing at the single-device level, which acts as a current bottleneck for the development of organic-based complex neuromorphic hardware systems. We recognize this issue, and present here a detailed analysis of one specific neuromorphic functionality, namely the short-term plasticity (STP) in flexible OFET synaptic devices. By combining experimental measurements and numerical modeling, systematic understanding of the voltage-dependent transmission behavior at the synaptic junction is obtained. By increasing the input-spike voltage magnitude, slowing down of both charging and discharging is observed, as the floating carrier reservoir turns electrostatically populated. The detailed analysis from this study builds a solid foundation for advanced models and the realization of flexible organic neuromorphic circuitries.

2. Experimental Methods

The organic synaptic transistors based on a floating-gate OFET architecture were fabricated with dinaphtho[2,3-b:2′,3′-f]thieno[3,2-b]thiophene (DNTT) semiconductor (Figure 1a), according to the bottom-gate, top-contact structure depicted in Figure 1b. A key to this device is the utilization of the ultra-thin, flat Al nanolayer, which is surface-oxidized to form an Al/AlOx stack [16]. The device fabrication processes are summarized as follows. The gate substrate is prepared as a flexible and transparent polyethylene terephthalate (PET) sheet, which has a predeposited conducting indium tin oxide (ITO) film (130 nm) (surface resistivity 60 Ω/sq, Sigma-Aldrich). The ITO surface was planarized by a 40-nm thick poly(3,4-ethylenedioxythiophene):poly(styrenesulfonate) (PEDOT:PSS, Clevios™, Heraeus) buffer layer to reduce the gate leakage. Then, insulating poly(methyl methacrylate) (PMMA, M.W. = 120,000, Sigma-Aldrich) was spin-coated from a toluene solution to serve as a blocking dielectric (410 nm). The Al functional layer with a nominal thickness of 3 nm was thermally evaporated and exposed to ambient air for oxidation. DNTT (sublimed grade, 99%, Sigma-Aldrich) was vacuum-evaporated for a hole-transporting molecular channel (50 nm). Finally, the Au source/drain electrodes (30 nm) were evaporated through a shadow mask.

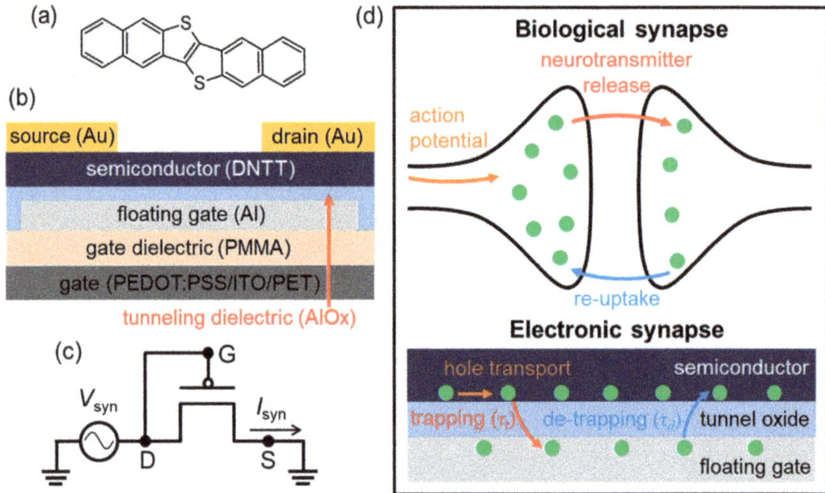

Figure 1. (**a**) Chemical structure of dinaphtho[2,3-b:2′,3′-f]thieno[3,2-b]thiophene (DNTT), used as an active molecular material in organic synaptic transistors. (**b**) Cross-sectional illustration of the device structure. (**c**) Circuit diagram employed for the measurement of voltage-dependent synaptic plasticity. (**d**) Model illustration: Biological processes relevant to the neuronal signal transmission through release and re-uptake of neurotransmitters and the electronic processes that mimic such properties through trapping-mediated hole transport at the semiconductor channel.

For emulating the STP behavior, we used the quasi-two-terminal electrical configuration shown in Figure 1c. Here, the gate and drain electrodes were externally wired and connected to a common computer controlled source-measure unit (Keithely 2400). The synaptic voltage (V_{syn}) pulses were generated by using a LabVIEW code, for them to have a specific number of sharp spike-like electrical stimulation stages with varying frequencies. The synaptic current (I_{syn}) was measured as a function of time as the output signal.

3. Results and Discussion

Materials characterization including atomic force microscopy (AFM) and transmission electron microscopy (TEM) as well as the basic transistor characterization such as transfer, output, mobility measurements has been reported in our previous paper [16]. Here, we introduce a numerical model that is applied to analyze the experimental STP. It is based on the functional model initially developed for nanoparticle organic memory field-effect transistors (NOMFET), by Bichler and co-workers [21]. In this study, we modify the notations and introduce the correlative parametrization approach, so that it can better describe our synaptic devices. Let's first recapitulate the physical meaning of such a model, by drawing a parallel between a biological synapse and a synaptic transistor channel. As shown in Figure 1d, the communication through a biological synapse is signaled by the action potential of the presynaptic neuron, which in turn activates the release of neurotransmitters toward the postsynaptic neuron. Part of these chemical messengers are eventually pulled back into the presynaptic neuron through re-uptake. Therefore, a specific time-domain pattern of neural signals conditions the dynamic variation of synaptic strength. Similarly, in our OFETs, holes accumulated at the DNTT channel serve as electronic signal carriers. Since the V_{syn} is bound to drain and gate, part of the transporting holes are trapped into the Al floating gate when a spike arrives. These trapped carriers can be easily detrapped into the channel, which is a key feature of our transistors with an ultra-thin tunnel oxide. Therefore, the STP behavior can be emulated by adjusting the input V_{syn} pulses.

For the square-type input V_{syn} waveform consisting of varying frequency and duty cycles, the direct relationship between the nth synaptic current I_n and the $(n + 1)$th one I_{n+1} can be iteratively established. As an intermediate, the current I_{n+} is the value at the falling edge of each synaptic spike and is dictated by how much the floating gate is charged during that pulse, which is written as

$$I_{n+} = I_n \exp\left(-\frac{W}{\tau_t}\right) + I_0\left[\exp\left(-\frac{\Delta E_F}{kT}\right)\left\{1 - \exp\left(-\frac{W}{\tau_t}\right)\right\}\right] \quad (1)$$

where W is the activation pulse width, τ_t is the trapping time constant, I_0 is the initial current, ΔE_F is the semiconductor Fermi-level shift at the fully charged state of the floating gate, k is the Boltzmann constant, and T is the absolute temperature. Between two pulses (while V_{syn} = 0 V), the carriers now leave the floating gate by natural detrapping, partially recovering the channel current, expressed as

$$I_{n+1} = I_{n+} \exp\left(-\frac{T_p - W}{\tau_d}\right) + I_0\left[1 - \exp\left(-\frac{T_p - W}{\tau_d}\right)\right] \quad (2)$$

where T_p is the pulse time period and τ_d is the detrapping time constant. Merging Equations (1) and (2) gives the final model

$$I_{n+1} = I_n \exp\left(-\tfrac{W}{\tau_t}\right)\exp\left(-\tfrac{T_p-W}{\tau_d}\right) + I_0\left[\exp\left(-\tfrac{\Delta E_F}{kT}\right)\left\{1 - \exp\left(-\tfrac{W}{\tau_t}\right)\right\}\exp\left(-\tfrac{T_p-W}{\tau_d}\right) + \left\{1 - \exp\left(-\tfrac{T_p-W}{\tau_d}\right)\right\}\right]. \quad (3)$$

To gain insights into the voltage-dependent signal transmission properties, we experimentally recorded the STP behavior of the same transistor, at four different magnitudes of V_{syn} as −4, −6, −8, and −10 V. The composition of the input signals (i.e., the frequency sequence and the number of spikes at each stage) was kept the same except for the voltage magnitude. Our test input waveforms consisted of six stages with the frequencies of 5, 1, 0.2, 2, 4, and 0.5 Hz. These frequencies determine the value of T_p, and W was fixed as 20 ms. Therefore, the remaining task in modeling is to fit the experimental I_{syn} data by determining four parameters, which are I_0, ΔE_F, τ_t, and τ_d. Instead of setting all these fitting parameters free, we employed a correlative extraction approach for more physically reliable results. The main idea is that the asymptotic final current $I_0\exp(-\Delta E_F/kT)$ should reflect the same amount of trapped carriers, and therefore have a quadratic dependence on the V_{syn} magnitude considering the forced saturation-regime transistor operation. To systematically apply this method, we first extracted the four fitting parameters from the data set at the lowest value of V_{syn} = −4 V. Then, we calculated the $I_0\exp(-\Delta E_F/kT)$ value for V_{syn} = −4 V, and then let this base asymptotic limit quadratically increase with increasing V_{syn}. Therefore, for the three other data sets (V_{syn} = −6, −8, and −10 V), the apparent initial I_0 value together with the prefixed $I_0\exp(-\Delta E_F/kT)$ value allowed for the unambiguous calculation of ΔE_F for each V_{syn}.

Figure 2a shows that I_0 monotonously increases in magnitude with increasing V_{syn} values, which is accounted for by the channel current flow enhanced by both gate (free carrier density) and drain voltages (lateral electric field) [18]. Interestingly, the ΔE_F follows a similar trend before experiencing a small drop at a high V_{syn}. This evidences that gate-induced trapping (decreasing the free carriers) and gate-enhanced hole accumulation act together to set the right balance for the Fermi level approachable at the fully charged state [22]. The inset of Figure 2a confirms that the magnitude of V_{syn} and the asymptotic synaptic current follows the quadratic dependence, evidenced by slope 2 on this log-log representation.

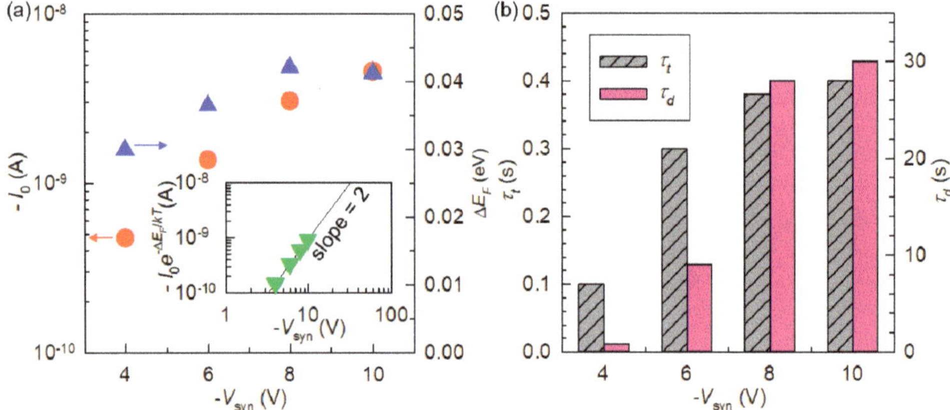

Figure 2. (**a**) The change of I_0 and ΔE_F as a function of voltage, estimated by the correlative approaches. Inset: the log-log plot showing the quadratic dependence of the final current on the synaptic voltage. (**b**) The extracted time constants for the trapping and detrapping processes.

Having determined the values of I_0 and ΔE_F, the two timing constants were estimated by performing global fitting to the experimental results. Figure 2b shows that despite the ultra-thin nature of our AlOx favoring spontaneous relaxation, the detrapping time constant τ_d is greater than the trapping counterpart τ_t at all voltage biases considered. Another important finding here is that the magnitude of V_{syn} can substantially influence the ratio between τ_d and τ_t values, implying a direct impact on the STP modulation.

Figure 3 shows the direct comparison between the experimental STP results and the model currents reproduced by inserting the parameters in Figure 2 into Equation (3). Similar STP behaviors have been observed in several field-effect synaptic transistors [14–16]. In brief, we can notice that even with the constant magnitude of V_{syn}, the produced I_{syn} quite significantly changes its magnitude responding to the spiking frequency. At a high V_{syn} frequency, a monotonous decrease in current is monitored because the negative gate pulse traps holes from the channel into the floating gate. When this frequency decreases, the amount of holes escaping the traps (per time) can exceed that of the holes being trapped into the floating gate, so that the I_{syn} gradually recovers its strength. In Figure 3, the model-calculated values are in a broad agreement with the measurements, and showed a similar trend in STP modulation. With increasing V_{syn}, the overall magnitude of output current I_{syn} went up, and it was necessary to introduce different timing parameters at each test voltage to fully explain the voltage-dependent transmission behavior. As shown in Figure 2b, the evolution of τ_d was more dramatic than that of τ_t, which is reflected in Figure 3 as the suppressed potentiation at $V_{syn} = -8$ or -10 V. This result also indicates that further optimization in synaptic voltages or structural engineering of nanoscale trapping media [23] may enable a switchable short-term and long-term neuromorphic behavior out of the same base architecture.

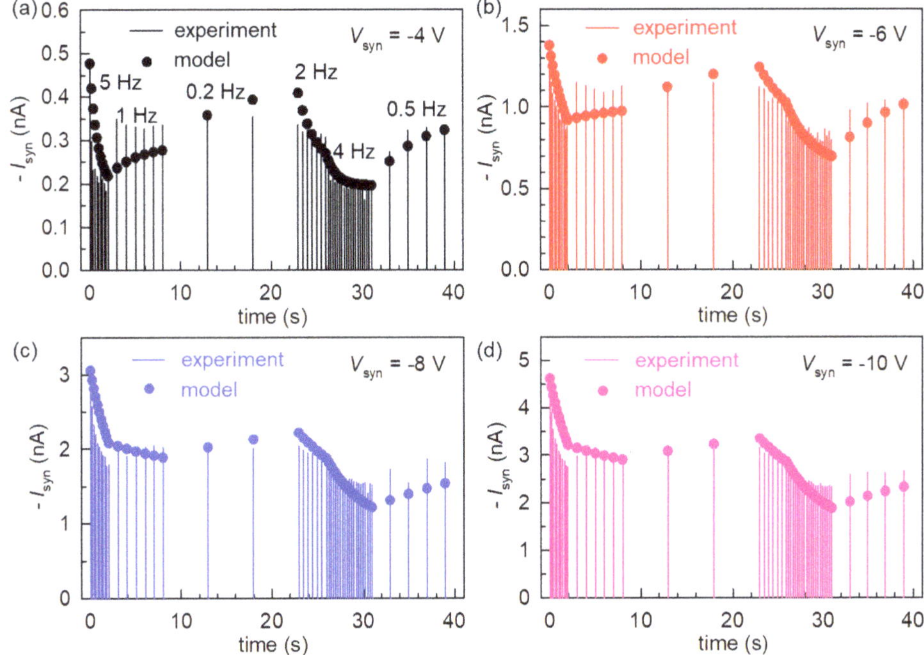

Figure 3. Comparing the experimental and model-reproduced STP (short-term plasticity) in organic synaptic transistors, with the magnitude of V_{syn} being (**a**) −4 V, (**b**) −6 V, (**c**) −8 V, and (**d**) −10 V. The test measurements consisted of six steps, the frequencies of which are denoted in (a). The same test condition applies to all the other panels.

4. Conclusions

We have reported on a combined experimental and theoretical analysis of the voltage-dependent synaptic plasticity in flexible OFETs. An iterative model was used in conjunction with the correlative extraction to understand the STP characteristics at different voltages. It was found that the applied voltage has a significant impact on I_0, ΔE_F, and timing constants. Among them, the τ_d experienced a particularly remarkable rise, turning the device into a practically depressing synapse at large voltages. At the same time, we have noticed the limited applicability of the model, evidenced by fitting errors. This indicates that an advanced model will need to be developed based on the physical characteristics of each complex trapping mechanism, which may for instance include the multiple time constants with a direct functional link to the materials and operational conditions.

Author Contributions: Conceptualization, S.L. and C.-H.K.; methodology, C.-H.K.; validation, S.L. and C.-H.K.; formal analysis, S.L. and C.-H.K.; investigation, S.L. and C.-H.K.; writing—original draft preparation, C.-H.K.; writing—review and editing, S.L. and C.-H.K. All authors have read and agreed to the published version of the manuscript.

Funding: This work was supported by the Gachon University research fund of 2019 (GCU-2019-0361), and also by the National Research Foundation of Korea (NRF) grant funded by the Korean government (MSIT) (NRF-2019R1C1C1003356).

Conflicts of Interest: The authors declare no conflict of interest.

References

1. Indiveri, G.; Liu, S.-C. Memory and information processing in neuromorphic systems. *Proc. IEEE* **2014**, *103*, 1379–1397. [CrossRef]

2. Dalgaty, T.; Payvand, M.; Moro, F.; Ly, D.R.B.; Pebay-Peyroula, F.; Casas, J.; Indiveri, G.; Vianello, E. Hybrid neuromorphic circuits exploiting non-conventional properties of RRAM for massively parallel local plasticity mechanisms. *APL Mater.* **2019**, *7*, 081125. [CrossRef]
3. Xia, Q.; Yang, J.J. Memristive crossbar arrays for brain-inspired computing. *Nat. Mater.* **2019**, *18*, 309–323. [CrossRef] [PubMed]
4. Upadhyay, N.K.; Jiang, H.; Wang, Z.; Asapu, S.; Xia, Q.; Yang, J.J. Emerging memory devices for neuromorphic computing. *Adv. Mater. Technol.* **2019**, *4*, 1800589. [CrossRef]
5. Mulaosmanovic, H.; Chicca, E.; Bertele, M.; Mikolajick, T.; Slesazeck, S. Mimicking biological neurons with a nanoscale ferroelectric transistor. *Nanoscale* **2018**, *10*, 21755–21763. [CrossRef]
6. Slesazeck, S.; Mikolajick, T. Nanoscale resistive switching memory devices: A review. *Nanotechnology* **2019**, *30*, 352003. [CrossRef]
7. Sun, J.; Fu, Y.; Wan, Q. Organic synaptic devices for neuromorphic systems. *J. Phys. D-Appl. Phys.* **2018**, *51*, 314004. [CrossRef]
8. Pecqueur, S.; Vuillaume, D.; Alibart, F. Perspective: Organic electronic materials and devices for neuromorphic engineering. *J. Appl. Phys.* **2018**, *124*, 151902. [CrossRef]
9. Van de Burgt, Y.; Melianas, A.; Keene, S.T.; Malliaras, G.; Salleo, A. Organic electronics for neuromorphic computing. *Nat. Electron.* **2018**, *1*, 386–397. [CrossRef]
10. Park, H.-L.; Lee, Y.; Kim, N.; Seo, D.-G.; Go, G.-T.; Lee, T.-W. Flexible neuromorphic electronics for computing, soft robotics, and neuroprosthetics. *Adv. Mater.* **2019**, *32*, 1903558. [CrossRef]
11. Gkoupidenis, P.; Schaefer, N.; Garlan, B.; Malliaras, G.G. Neuromorphic functions in PEDOT:PSS organic electrochemical transistors. *Adv. Mater.* **2015**, *27*, 7176–7180. [CrossRef] [PubMed]
12. Van de Burgt, Y.; Lubberman, E.; Fuller, E.J.; Keene, S.T.; Faria, G.C.; Agarwal, S.; Marinella, M.J.; Talin, A.A.; Salleo, A. A non-volatile organic electrochemical device as a low-voltage artificial synapse for neuromorphic computing. *Nat. Mater.* **2017**, *16*, 414–418. [CrossRef] [PubMed]
13. Kim, C.-H. Self-rectifying DNTT memristors. *IEEE Electron. Device Lett.* **2018**, *39*, 1736–1739. [CrossRef]
14. Alibart, F.; Pleutin, S.; Guérin, D.; Novembre, C.; Lenfant, S.; Lmimouni, K.; Gamrat, C.; Vuillaume, D. An organic nanoparticle transistor behaving as a biological spiking synapse. *Adv. Funct. Mater.* **2010**, *20*, 330–337. [CrossRef]
15. Hafsi, B.; Boubaker, A.; Guerin, D.; Lenfant, S.; Desbief, S.; Alibart, F.; Kalboussi, A.; Vuillaume, D.; Lmimouni, K. Electron-transport polymeric gold nanoparticles memory device, artificial synapse for neuromorphic applications. *Org. Electron.* **2017**, *50*, 499–506. [CrossRef]
16. Kim, C.-H.; Sung, S.; Yoon, M.-H. Synaptic organic transistors with a vacuum-deposited charge-trapping nanosheet. *Sci. Rep.* **2016**, *6*, 33355. [CrossRef] [PubMed]
17. Gelinck, G.; Heremans, P.; Nomoto, K.; Anthopoulos, T.D. Organic transistors in optical displays and microelectronic applications. *Adv. Mater.* **2010**, *22*, 3778–3798. [CrossRef] [PubMed]
18. Kim, C.-H.; Bonnassieux, Y.; Horowitz, G. Compact DC modeling of organic field-effect transistors: Review and perspectives. *IEEE Trans. Electron Devices* **2014**, *61*, 278–287. [CrossRef]
19. Guo, X.; Xu, Y.; Ogier, S.; Ng, T.N.; Caironi, M.; Perinot, A.; Li, L.; Zhao, J.; Tang, W.; Sporea, R.A.; et al. Current status and opportunities of organic thin-film transistor technologies. *IEEE Trans. Electron Devices* **2017**, *64*, 1906–1921. [CrossRef]
20. Yu, K.; Park, B.; Kim, G.; Kim, C.-H.; Park, S.; Kim, J.; Jung, S.; Jeong, S.; Kwon, S.; Kang, H.; et al. Optically transparent semiconducting polymer nanonetwork for flexible and transparent electronics. *Proc. Natl. Acad. Sci. USA* **2016**, *113*, 14261–14266. [CrossRef]
21. Bichler, O.; Zhao, W.; Alibart, F.; Pleutin, S.; Vuillaume, D.; Gamrat, C. Functional model of a nanoparticle organic memory transistor for use as a spiking synapse. *IEEE Trans. Electron Devices* **2010**, *57*, 3115–3122. [CrossRef]
22. Kim, C.H.; Bonnassieux, Y.; Horowitz, G. Fundamental benefits of the staggered geometry for organic field-effect transistors. *IEEE Electron Device Lett.* **2011**, *32*, 1302–1304. [CrossRef]
23. Kim, C.-H. Nanotrapping memories. *Nanoscale Horiz.* **2019**, *4*, 828–839. [CrossRef]

Article

A Spiking Neural Network Based on the Model of VO_2–Neuron

Maksim Belyaev * and Andrei Velichko

Institute of Physics and Technology, Petrozavodsk State University, 31 Lenina str., Petrozavodsk 185910, Russia; velichko@petrsu.ru

* Correspondence: biomax89@yandex.ru

Received: 27 August 2019; Accepted: 18 September 2019; Published: 20 September 2019

Abstract: In this paper, we present an electrical circuit of a leaky integrate-and-fire neuron with one VO_2 switch, which models the properties of biological neurons. Based on VO_2 neurons, a two-layer spiking neural network consisting of nine input and three output neurons is modeled in the SPICE simulator. The network contains excitatory and inhibitory couplings, and implements the winner-takes-all principle in pattern recognition. Using a supervised Spike-Timing-Dependent Plasticity training method and a timing method of information coding, the network was trained to recognize three patterns with dimensions of 3×3 pixels. The neural network is able to recognize up to 10^5 images per second, and has the potential to increase the recognition speed further.

Keywords: leaky integrate-and-fire neuron; vanadium dioxide; neural network; pattern recognition

1. Introduction

Artificial neural networks (ANNs), created by analogy with biological neural systems, are used to resolve various tasks, such as classification, clustering, and pattern recognition [1–4]. The main element of ANN is a neuron that may have several inputs and one output, and neurons can be connected in different ways, depending on the network architecture [5,6]. The main task of a neuron is to convert input signals to output signal using an activation function [5]. In the history of ANN, three generations of the networks are usually distinguished. The first generation includes simple forward and backward connection networks that operate with binary data and stepwise activation functions [7]. The second generation includes multilayer networks of direct and reverse distribution, operating with rational numbers with continuous activation functions [7]. The third generation of ANN (spiking neural networks (SNN)) uses biosimilar models of neurons that take into account not only the magnitude of the signals arriving at the input, but also the signals' temporal distribution [7,8].

There is a large number of SNNs, which are used to solve practical tasks and are based on mathematical models of neurons (Integrate-and-Fire, Izhikevich, Hodgkin-Huxley) [9–12]. Such SNNs use the resources of computers, video cards, and field-programmable gate arrays to emulate the network operation [9–13]. Although such SNNs currently provide impressive performance results [14], any emulation loses hardware implementation in performance and energy efficiency [15,16]. Therefore, the development of SNNs based on microelectronic elements attracts the active attention of researchers [16–22]. One of the most frequently used functional elements of SNN is a memristor [16], which is used for implementing customizable weights and as a functional element of a neuron. The weights are adjusted during the network training, and, in electric networks, it is implemented by changing the impedance of the lines connecting the outputs and inputs of neurons. A multi-stable resistive memory cell is an ideal object for implementing a wide SNN functionality, due to the possibility of changing the resistance over a wide range of values. However, the application of a resistive memory cell as a bi-stable element with an off state (inactive neuron) and on state (active neuron) is not an optimal solution because of the high probability of its resistance modification during the operation [23,24].

In the current study, for the manufacturing of artificial neurons, we propose to use elements with a stable S-shaped I – V characteristic, such as switches based on transition metal oxides with a metal-insulator transition [25–27]. Implementations of neuron models on the VO_2 switch are described in References [28–34]. However, a few SNN implementations using such neurons have been proposed so far. Models of VO_2 neurons can be divided into two groups. The first group is an integrate-and-fire model of a neuron [32,33,35], which has three main states: the accumulation of action potential state due to charging the capacitor, the spike generation state, when the capacitor is discharged, and the VO_2 switch goes into a highly conductive state, and the inactivity state of a neuron. The discharge time of the capacitor is treated as a post firing refractory period [30,32], and the initiation of the second pulse is impossible at that time due to shunting of the low resistance of the switch. The second group of models covers neuron circuits that include inductance, and the possibility of generating a burst mode, which is similar to the FitzHugh-Nagumo and FitzHugh-Rinzel models [34,35].

We propose a leaky integrate-and-fire (LIF) circuit for a neuron based on a VO_2 switch that can implement excitatory and inhibitory couplings. Based on VO_2 neurons, in the SPICE simulator, the operation of a two-layer SNN network consisting of nine input and three output neurons was modeled. An image in the form of a 3×3 matrix is fed to the network input, and, at the output, one of the three neurons is activated with a certain input pattern, and this neuron suppresses the remaining output neurons according to the winner-take-all (WTA) principle [36]. The coding of information in the proposed network is performed by setting the delay time of the spikes in the input layer relative to the zero time moment (time to the first spike) [37]. Network training is performed according to the spike time-dependent plasticity (STDP) scheme [14,16,19–21,38]. As a result, a model SNN, based on VO_2 neurons, which allows pattern recognition, is presented and investigated in this study.

2. SNN Modeling Method

2.1. VO_2 Neuron Model

The VO_2-neuron model is created on the basis of the LIF neuron model, which is widely used due to the simplicity of implementation and the possibility of generating biosimilar spikes [39]. Its main element is a bi-stable two-electrode VO_2 switch [25–27]. The operation principle of the switch is based on the metal-insulator phase transition in VO_2 films, which happens near the transition temperature $T_{th} \sim 340$ K. The critical temperature T_{th} in the film is achieved due to the Joule heating effect when passing a current, which leads to a sharp abrupt change in the resistance [27]. In addition to the thermal effect, when modeling electric switching, the effect of the electric field on the concentration of charge carriers is taken into account [40,41]. The model I – V characteristic of the VO_2 switch corresponds to the experimental I – V characteristic (Figure 1), measured in our previous work [25], on a planar switch with a channel size of 2.5–3 μm, and a VO_2 film thickness of ~ 250 nm, with a current limiting resistor of 250 Ω connected in series.

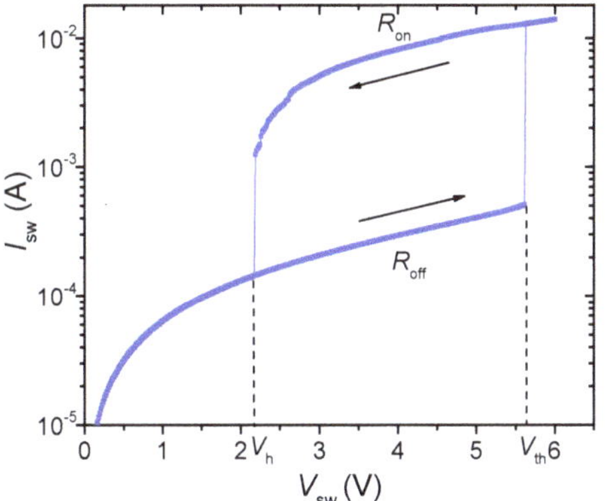

Figure 1. Experimental I–V characteristic of a planar VO_2 switch.

Figure 1 demonstrates the dependence of the switch current I_{sw} on the voltage V_{sw} supplied to the switch. Reaching the threshold switching voltages V_{th} = 5.6 V and holding voltage V_h = 2.2 V, the switch passes from a high-resistance state to a low-resistance state and vice versa. The high-resistance and low-resistance branches of the I – V characteristic are approximated by linear dependencies on the voltage V_{sw} with resistance values R_{off} ~ 14 kΩ and R_{on} ~ 300 Ω, respectively.

To conduct SPICE simulations of the VO_2 neuron, a standard voltage-controlled switch was used with parameters corresponding to the experimental I – V characteristics (R_{off}, R_{on}, $V_{th,}$ and V_h).

The electrical circuit of the VO_2 neuron is shown in Figure 2. The neuron model has n inputs, and one output V_{out}. Resistances $R_w^1 \ldots R_w^n$ play the role of a synaptic weights between neurons. The smaller the resistance, the more the signal from the *i*-th input affects the neuron. The spikes coming from the inputs through the resistances are accumulated on the C_{sum} capacitance, by charging it with the cumulative charge. The charge from the C_{sum} capacitor gradually flows through the resistance R_{in}. C_{sum} capacitance voltage is an effective input signal that affects the current state of a neuron. The supply voltage V_{dd} is selected so that the VO_2 switch stays in the off state in the absence of input signals. The most clear way to achieve this condition is to set the voltage V_{dd} less than the switching voltage V_{th}. In this model, the inactive state of the neuron corresponds to the switched off VO_2 switch when it is in the subthreshold mode ($V_{sw} < V_{th}$). To activate the neuron, the switch should be turned on by setting the voltage on the switch to $V_{sw} \geq V_{th}$. To achieve this, the supply voltage V_{dd} must have negative values, and the spikes supplied to the input must have a positive polarity.

To activate a neuron, the voltage across the capacitance C_{sum} should increase to a threshold value V_{c_th}, which depends on V_{dd}, the resistance of the switch in the off state R_{off}, and the values of the resistors R_s and R_{in}. After the switch is turned on, its resistance decreases to R_{on}, which leads to the discharge of the capacitance C_c through the resistances R_{in} and R_{out}. The capacitance C_c serves as a reservoir of charge, which is necessary for generating a spike when a neuron is activated. When C_c is discharged, a spike of positive polarity is generated at the output V_{out} of the neuron. By connecting the outputs of some neurons with the inputs of other neurons, SNNs with excitatory coupling can be obtained. The resistance R_s is load resistance and sets the operating current through the switch. R_s is selected in a way that the VO_2 switch turns off after discharging the capacitance C_c. In fact, the neuron circuit is tuned to generate a single current spike through the VO_2 switch, i.e., generate a single spike at the output.

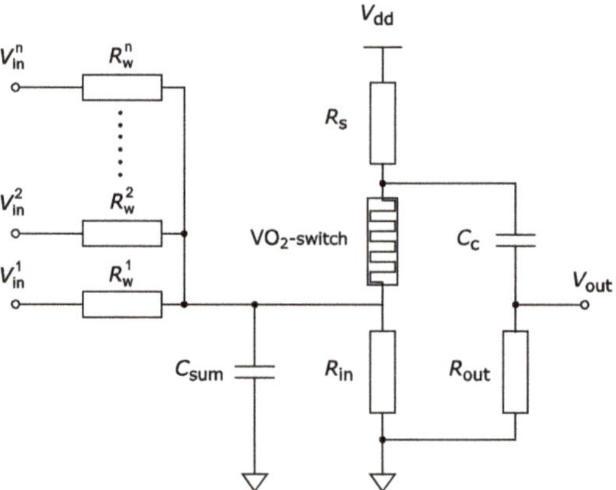

Figure 2. Electrical circuit of a VO_2 neuron.

Figure 3a presents the electrical circuit of a VO_2 neuron. The pulses from the voltage generator are supplied to the input of the neuron, and the output is connected to the input stage of the subsequent neuron (to simulate the output load of the neuron). The circuit modeling was performed in the LTspice XVII simulation software. Resistance and capacitance values: $R_{w\_1} = 500\ \Omega$, $R_{w\_2} = 1\ k\Omega$, $R_s = 700\ \Omega$, $R_{in} = 1\ k\Omega$, $R_{out} = 10\ k\Omega$, $C_{sum} = 1$ nF, and $C_c = 10$ nF. Supply voltage $V_{dd} = -5.75$ V. A pulse of positive polarity with an amplitude of 2 V and a duration of 0.3 μs is supplied from the generator. Figure 3b depicts the oscillograms of the input V_{in} and output V_{out} voltages, as well as the voltages V_c at the capacitance C_{sum}, which demonstrates the spikes' dynamics.

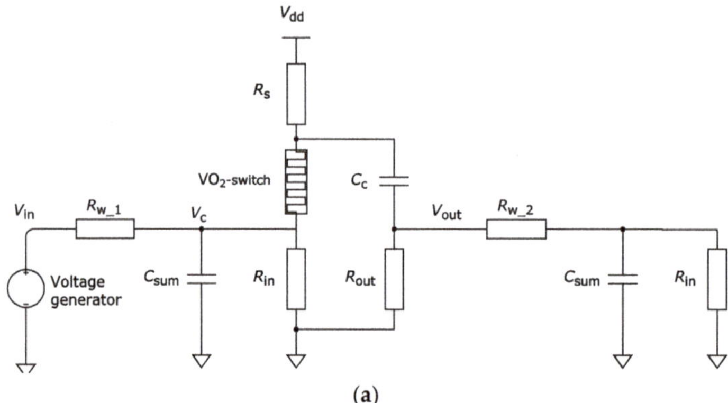

(a)

Figure 3. *Cont.*

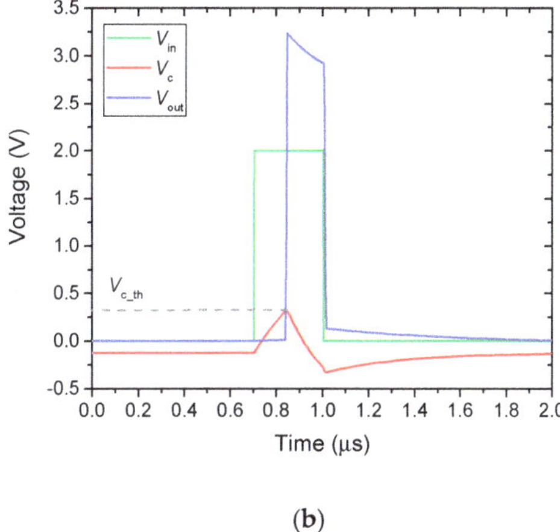

(b)

Figure 3. (**a**) An example of an electrical circuit of a VO_2 neuron activated by a voltage generator, and (**b**) oscillograms of voltages V_{in}, V_{out}, and V_c illustrating the spikes' dynamics.

The threshold voltage at the C_{sum} capacitance, required to initiate the output spike, is V_{c_th} ~ 0.33 V (dashed line in Figure 3b). After turning on the VO_2 switch, the capacitance C_c starts to discharge, and it leads to the appearance of a leading edge and a spike with a voltage amplitude of ~ 3.2 V. After turning on the switch, a decrease in the voltage V_c to negative values is associated with active recharging of the C_{sum} capacitor through an open switch due to the negative voltage on C_c capacity and V_{dd} power supply. The spike duration is ~ 170 ns, which is determined by the discharge time of the capacitance C_c until the moment, when the voltage at the switch V_{sw} is not less than V_h. The trailing edge of the pulse appears when the switch goes off. The duration of the output spike can be significantly longer than the duration of the initiating pulse.

The VO_2-neuron model is able to demonstrate various properties of real neurons, such as spike latency, subthreshold oscillations, refractory period, threshold behavior, and spike frequency adaptation [28,39].

For example, Figure 4a demonstrates that the higher the amplitude of the input pulse exists, the smaller the time delay between the leading edges of the input and output pulses remains, called spike latency. With a pulse amplitude of 2 V, the latency between the input and output signals is 140 ns, and with a pulse amplitude of 1 V, the latency reaches 440 ns. Therefore, the amplitude and duration of the input pulse, required to initiate the spike, can lie in a wide range. However, when the amplitude of the input pulse is less than V_{c_th}, the initiation of the output pulse does not occur.

If the input pulse is sufficiently long, several spikes can be obtained at the output of the circuit. Figure 4b demonstrates the response of a VO_2 neuron to a pulse with an amplitude of 1 V and a duration of 3.6 μs, which forms five spikes at the output. The latter mode resembles the occurrence of oscillations when an excitation signal is applied (subthreshold oscillations). The delay between the spikes T_r, called the refractory period, is approximately 630 ns and is determined by the charging time of the capacitor C_{sum} to voltage V_{c_th}. The refractory period depends on the amplitude of the pulse. For example, at a pulse amplitude of 2 V, the period T_r is 300 ns. In addition, the refractory period T_r is slightly increasing (see the values indicated in Figure 4b), because of the small increase in V_{c_th} from spike to spike, since the capacitance C_c does not have time to charge to its original values. This increase in the time period between the spikes under constant exposure is similar to biological neurons (spike frequency adaptation) [28,39].

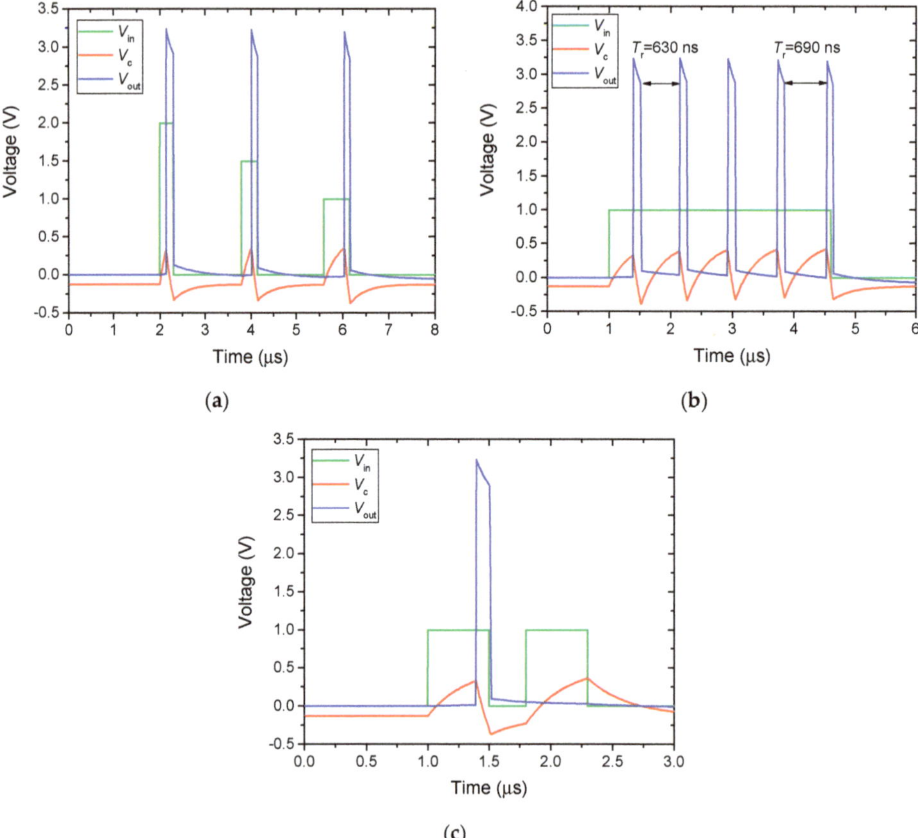

Figure 4. (**a**) Oscillograms of V_{in}, V_c, and V_{out}, applying to the VO_2-neuron input pulses of different duration and amplitude, (**b**) one long pulse, and (**c**) two pulses with a small delay between them.

If the delay between the spikes is less than the refractory period, the neuron generates only one spike. Figure 4c demonstrates two input pulses with an amplitude of 1 V and a delay of 300 ns, and the neuron generates a spike only for the first input pulse.

To implement the wide functionality of neural networks, in addition to excitation connections, the possibility to add inhibitory connections is required. Inhibitory connections are widely used in the SNN output layer to implement the WTA rule. Such connections allow the first spike-generated neuron to deactivate all other related neurons using the inhibitory connections. As a result, only one neuron, which is associated with a recognized class, is activated. Figure 5a demonstrates a diagram of two neurons interconnected via capacitances C_{inh} = 10 nF, which act as inhibitory connections. The capacitance and resistance values correspond to the single neuron circuit shown in Figure 3a, with the exception of R_{in} = 200 Ω and R_{out} = 200 Ω. Due to the presence of C_{inh} capacities, upon activation of one of the neurons and the discharge of its capacitance C_c, the voltage on the capacitance C_c of an inactive neuron decreases. In this case, the first (in time) activated neuron will suppress all other neurons connected to it by inhibitory connections. Namely, in such a group of neurons, the WTA rule is implemented. In order to trace the activation of neurons in this circuit, it is convenient to monitor the current I_{sw} and voltage V_{sw} on two switches (Figure 5b). The delay between the supplied pulses V_{in_1} and V_{in_2} is 2 μs. When the first pulse V_{in_1} arrives at the first switch (Figure 5b), the switch turns on, the current I_{sw_1} increases sharply, and the on mode lasts for ~ 4.2 μs. Switching on occurs because the voltage V_{sw_1} reaches the threshold value V_{th} (Figure 5b).

After turning on the first switch, the voltage V_{sw_1} drops sharply and it leads to a decrease in voltage V_{sw_2} on the second switch, as the signal is transmitted through the capacitors C_{inh}. The second pulse arriving at the input of the second neuron (V_{in_2}) does not activate it, because the voltage V_{sw_2} does not reach the threshold value ($V_{sw\_2} < V_{th}$). The activation of the first neuron inhibits the activation of the second neuron. If an excitation pulse is applied to the second neuron after deactivation of the first neuron, then the second neuron will go into an active mode.

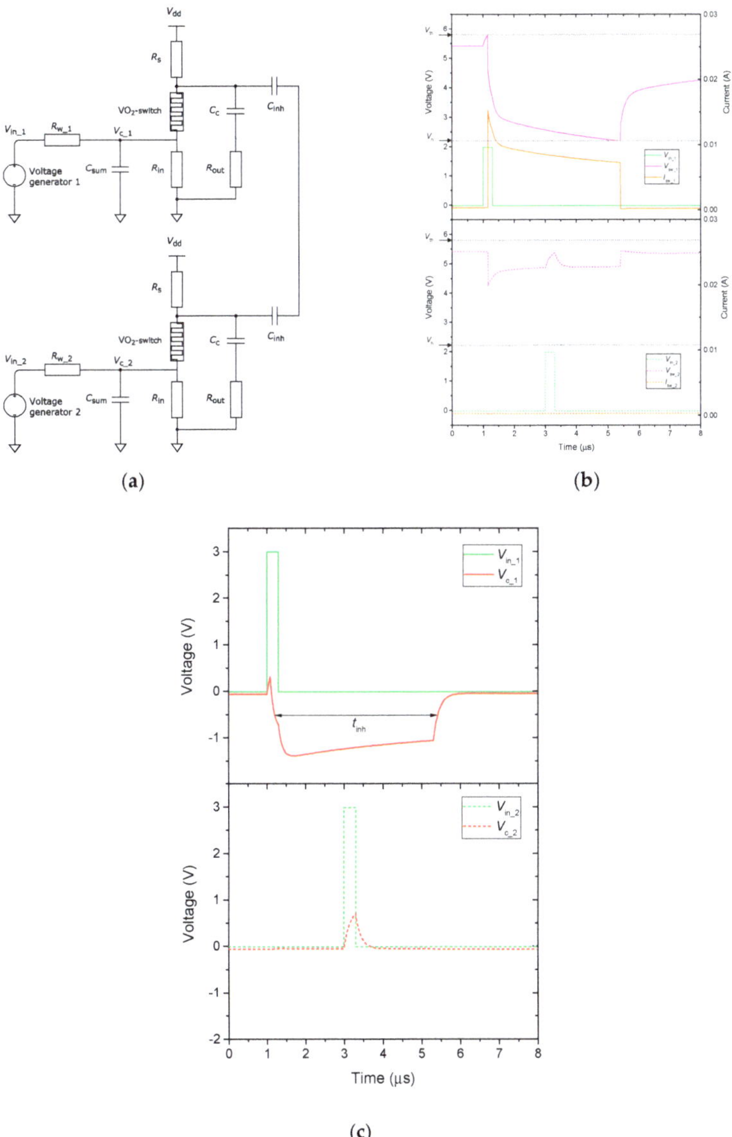

Figure 5. (**a**) Connection diagram of two oscillators with inhibitory connections. (**b**) Oscillograms of the input signals V_{in_1}, V_{in_2}, voltages V_{sw_1}, V_{sw_2} and currents I_{sw_1}, I_{sw_2} on the switches, and (**c**) oscillograms of the input signals V_{in_1}, V_{in_2} and voltages on the integrating capacitors V_{c_1}, V_{c_2}, when applying two voltage pulses with a delay of 2 µs.

To determine the activity of neurons in the output layer, the most appropriate solution would be to convert the I_{sw} current pulses into output voltage pulses. However, this solution requires additional external circuits. As neurons are connected by C_{inh} capacities, voltages, taken from R_{out} resistors, are correlated, and it is not advisable to use them. Schematically, as activity markers, the voltages V_{c_1} and V_{c_2} can be used, and their dynamics are shown in Figure 5c. When the first neuron is activated, the voltage V_{c_1} drops sharply due to the recharging of the capacitor C_{sum}, which forms a strong pulse of negative polarity, and the positive pulse V_{c_2} is weakly expressed on the inactive neuron.

2.2. SNN Architecture

For pattern recognition problems, various SNN architectures are used, which differ in the number of layers and in the way neurons are connected [14,16]. One of the simplest SNN architectures is a two-layer network (Figure 6a), where image information is supplied to the input (first layer), and one of the neurons associated with a certain class of images is activated at the output (second layer) [42–44]. Each of the first layer neurons is connected to each neuron of the second layer through excitatory connections. The connection strength between each pair of neurons is specified through synaptic weights, which can vary among themselves. All neurons of the output layer are interconnected by inhibitory connections. When applying signals to the first layer neurons, they are activated and transmit an excitation effect to the second layer. The neuron of the output layer, which is activated first, sends an inhibitory signal on all other output neurons. This prevents their activation. In this way, the WTA rule is implemented, when data is classified by defining the only active neuron in the output layer. The activation speed of the output layer neurons depends on the input signals and synaptic weights between the particular output neuron and each neuron from the input layer. For the correct pattern recognition, during the network training, it is necessary to correctly set the synaptic weights for each group of the output neuron on the input neurons.

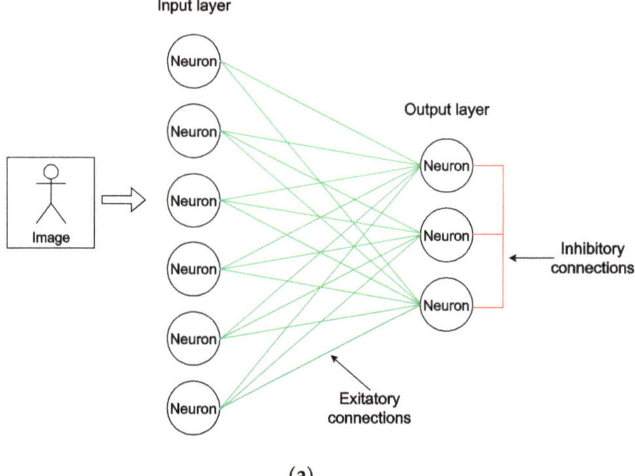

(a)

Figure 6. *Cont.*

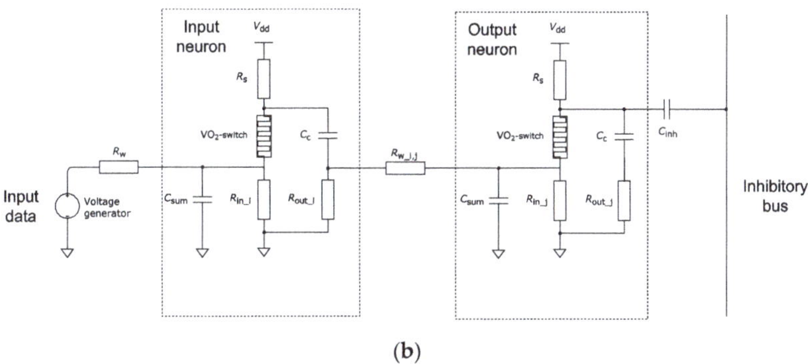

(b)

Figure 6. (**a**) Architecture of a two-layer neural network for pattern recognition and (**b**) circuit implementation of neurons in the input and output layers.

Figure 6b presents a coupling diagram of neurons of the input and output layers of a pulsed neural network for image classification. Each element of the input layer corresponds to one pixel of the image. Therefore, to solve the problem of classifying images with a size of 3 × 3 pixels, nine neurons in the input layer are required. The number of output neurons depends on the number of patterns that the network is supposed to recognize. In this study, we will demonstrate the classification of images using three patterns, so the number of neurons in the output layer will be three. The input and output layers of the SNN are connected using synaptic weights, implemented through the resistances $R_{w\_i,j}$, where i is the number of the input neuron and j is the number of the output neuron. The resistance values $R_{w\_i,j}$ will change during training. Memristors [16], where resistance can be adjusted, are often used as resistances in the circuits. In this study, we do not consider a circuit implementation that allows the change of resistances $R_{w\_i,j}$ during the training process. Instead, we assume to have control over the elements' resistances. The range of resistance values $R_{w\_i,j}$ varies from 1.5 kΩ to 2.5 kΩ.

A signal from the generator, which encodes information about the color of the pixel, is supplied to the input of each input layer neuron. In this study, we use eight-bit grayscale images, so the information encoded by the generator reflects a gray scale, where the black pixel corresponds to the number 0 and the white pixel corresponds to the number 255. The generator is connected to the input layer neuron using an excitation connection. Then, all nine neurons of the input layer are connected by excitation connections to the three neurons of the output layer, which forms 9 × 3 = 27 connections.

All output neurons are interconnected by inhibitory connections, which are implemented by connecting to the inhibitory bus using the capacitance C_{inh} = 10 nF.

The remaining elements, depicted in Figure 6b, have the following ratings: R_w = 500 Ω, R_s = 700 Ω, R_{in_i} = 1 kΩ, R_{in_j} = 200 Ω, R_{out_i} = 10 kΩ, R_{out_j} = 200 Ω, C_{sum} = 1 nF, and C_c = 10 nF. The supply voltage of all neurons is V_{dd} = −5.75 V.

2.3. SNN Training

Before considering the network training algorithm, it is necessary to determine the method of information coding. A large number of information coding methods for SNN has been defined: rate coding, rank coding, time to first spike, latency coding, phase coding, population coding, and others [37]. Typically, two-layer neural networks, used to classify images, apply the rate coding method [43,45,46]. However, in the current study, we use the time to the first spike method [37]. This coding method requires fewer spikes for a single recognition act, and, as a result, less energy is spent on the circuit operation, as most of the energy is spent on generating spikes.

The information coding is performed as follows. The signals from the generators arrive on the first layer of the neural network with a delay Δt relative to the start time of the circuit t = 0, and the delay Δt determines the brightness of the image pixel. The value Δt = 0 corresponds to brightness

0 (black color), and the maximum delay Δt_{max} = 2 μs corresponds to brightness 255 (white color). The signal from the generator is a rectangular pulse with an amplitude of 2 V and a duration of 0.3 μs. The delay time is counted relative to the leading edge of the pulse.

The network training process is based on the standard STDP mechanism [14,16,19–21,38]. This mechanism is an implementation of the Hebbian learning rule and causes a change in synaptic weight depending on the delay $\Delta t_{in\text{-}out}$ between pre-synaptic and post-synaptic spikes [45]. The traditional rule is an exponential function [45], which depends on $\Delta t_{in\text{-}out}$. However, various studies use the simplified versions [42,47], which significantly facilitate the calculations, while maintain the main ideas of the SPDT method. In this study, we use the function presented in Figure 7, where the form is given in the SNN training papers [42,45,47]. Since an increase in synaptic weight corresponds to a decrease in resistance $R_{w\_i,\,j}$, the function is inverted in relation to the axes of an ordinate. Resistance decreases, if the post-synaptic spike (from the neuron in the output layer) arrives with a delay in the range of 0 to 0.5 μs after the pre-synaptic spike (from the neuron in the input layer). In other cases, the resistance $R_{w\_i,\,j}$ increases.

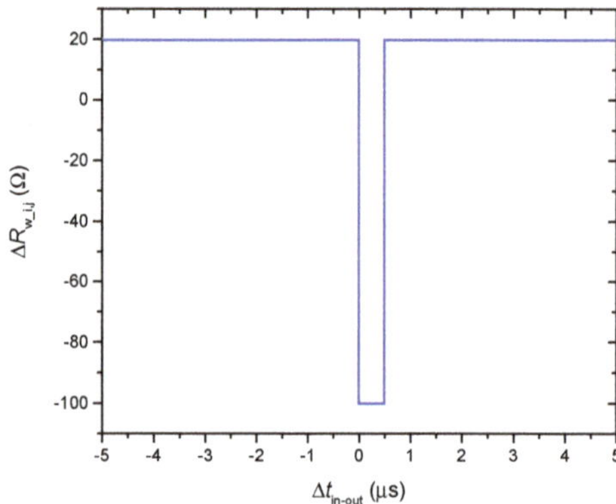

Figure 7. The function of the resistance change the between the input and output neuron $\Delta R_{w\_i,\,j}$ depending on the delay between the pre-synaptic and post-synaptic spikes $\Delta t_{in\text{-}out}$.

Typically, the STDP-based training procedure is used in SNN with unsupervised learning [19–21, 42,44]. In this case, when training input data is supplied, the output neurons are randomly associated with input data patterns. Nevertheless, there are studies on SNN training mechanisms that implement supervised learning [48–50]. In these studies, the input pattern is forcibly assigned to a specific output neuron using back error propagation algorithms. In our study, we tried to implement a simplified approach that allows us to implement supervised learning. During the training, the supply voltage V_{dd} was set to be non-zero only at one of the three output neurons (see Table 1). The remaining neurons are forcibly electrically deactivated. They do not emit spikes, and it causes all the associated weights $R_{w\_i,\,j}$ to increase. During network training (Table 1), power at the output neuron No. 1 is present ($V_{dd} \neq 0$) only when "Pattern 1" images are inputted. When "Pattern 2" and "Pattern 3" images are supplied, the voltage V_{dd} is zero.

Table 1. An example of the supply voltage setting V_{dd} of the output layer, using the supervised learning method in SNN training.

The Class of the Image, Fed to the SNN Input	The Voltage V_{dd} of the Output Neuron No. 1, V	The Voltage V_{dd} of the Output Neuron No. 2, V	The Voltage V_{dd} of the Output Neuron No. 3, V
Pattern 1	−5.75	0	0
Pattern 2	0	−5.75	0
Pattern 3	0	0	−5.75

The SNN training algorithm consists of the steps listed in Figure 8. First, arbitrary resistances $R_{w\_i,j}$ are set in the range from 1.5 kΩ to 2.5 kΩ. Second, the iterative process of changing the resistances $R_{w\_i,j}$ begins. Initially, one of the patterns, that the network should be trained to express, is arbitrarily selected (the number of patterns should be equal to the number of the output layer neurons). In accordance with the pattern and the information coding scheme, the pulse delays are set to be supplied from the generators to the input layer neurons. Then, in accordance with Table 1, the V_{dd} values of the output neurons are set. Next, the circuit modelling starts in the SPICE simulator. Based on the simulation results, delays between pre-synaptic and post-synaptic spikes $\Delta t_{in\text{-}out}$ are calculated, and $\Delta R_{w\_i,j}$ are calculated using the resistance change function (Figure 7). After that, the new values of $R_{w\_i,j}$ are set, and, if the values are outside the range of 1.5 kΩ – 2.5 kΩ, $R_{w\_i,j}$ is set equal to the nearest border value. The training cycle is repeated, until all the $R_{w\_i,j}$ values stop changing.

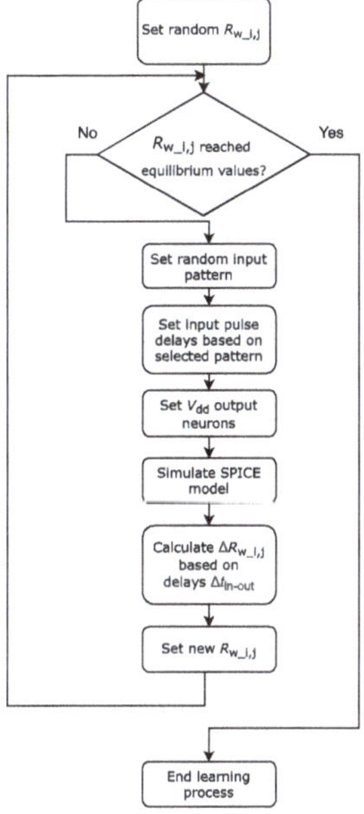

Figure 8. SNN training algorithm.

3. Results

Three patterns for training with a dimension of 3 × 3 pixels are presented in Figure 9. The patterns have the same number of black and white pixels. If the number of black pixels is different, then normalization by color intensity can be applied to obtain more accurate results [51].

Figure 9. Set of patterns used for SNN training.

Figure 10 illustrates the resistance values $R_{w\_i,1}$, $R_{w\_i,2}$, and $R_{w\_i,3}$ between all input neurons and three output neurons before and after network training. Resistance values are grouped by nine pieces according to the number of output neurons.

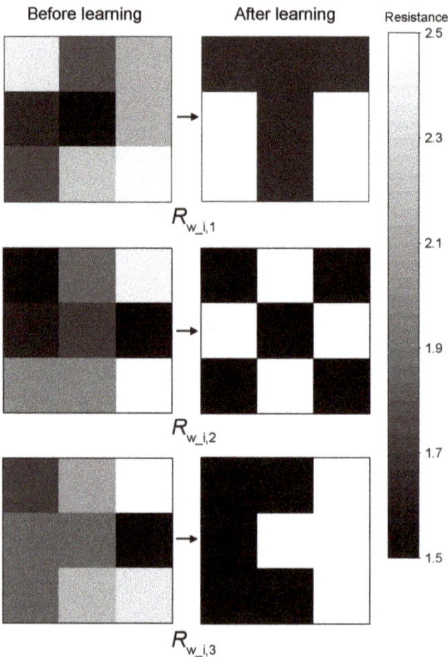

Figure 10. Distribution of resistances $R_{w\_i,1}$, $R_{w\_i,2}$, and $R_{w\_i,3}$ before and after training.

Before training, the resistances $R_{w\_i,j}$ were randomly generated in the range from 1.5 kΩ to 2.5 kΩ. Then, the SNN training procedure was performed for ~100 cycles, described in detail in Chapter 2.3, using the input patterns in Figure 9. As a result, the distribution of resistances for each output neuron began to correspond to the pattern assigned to each output neuron. This distribution is an expected result for two-layer networks operating, according to the WTA mechanism [44]. If the training patterns were a set of images distributed by classes (for example, numbers written in different handwriting), then the distribution of weights would be averaging all patterns belonging to the same class. Such a training outcome is observed in the studies using the MNIST database [42].

Analysis of the trained network reveals the following results. If patterns from the training set were input, then, as expected, the neurons corresponding to the associated pattern were activated at the SNN output.

Furthermore, images corresponding to distorted patterns in which pixel color intensities were randomly changed were inputted. The results of the SNN operation with the number of the activated output neuron are presented in Figure 11. The first three images corresponding to the distorted patterns from the training set were correctly classified by SNN. The first image corresponds to "Pattern 3," the second image corresponds to "Pattern 1," and the third image corresponds to "Pattern 2."

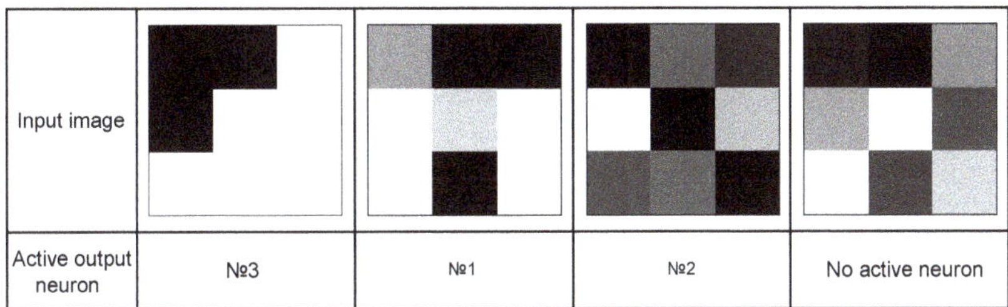

Figure 11. Examples of image classification of distorted patterns indicating activated output neurons.

When inputting the fourth image, which is a highly distorted template from the training set ("Pattern 3"), none of the output neurons were activated, and it could be interpreted as an undefined result. Inactivity of the output neurons reflects that the voltage on the capacitors C_{sum} of the output neurons never reaches the threshold values $V_c < V_{c\_th}$.

This uncertainty could be avoided by reducing the time range for spikes coding Δt_{max} (see Section 2.3). For example, if $\Delta t_{max} = 2$ µs is reduced by half to $\Delta t_{max} = 1$ µs, then the fourth image in Figure 11 is correctly identified as "Pattern 3." The pulses have shorter time intervals, and the voltage on the integrating capacitance C_{sum} is more likely to increase to the threshold value V_{c_th}.

4. Discussion

When modelling neurons, on the one hand, we could strive for greater bio-similarity of a neuron, as implemented in some models (FitzHugh-Nagumo, Izhikevich, Hodgkin-Huxley) [35,39], or, on the other hand, we could try to minimize the number of electrical components in the circuit to contribute to its miniaturization in practical implementation. The integrate-and-fire neuron model, which we propose, permits, on the one hand, to obtain a number of properties observed in biological systems (Section 2.1), and, on the other hand, it contains only one switching element and a power source, in contrast to more complex models [28]. In a number of studies of neurons based on VO_2, silicon semiconductor devices (field effect transistor, diode) are used as additional circuit elements [22,29]. This imposes certain restrictions on the compatibility of technological processes for manufacturing the silicon and non-silicon parts of the circuit. This drawback is avoided in the presented neuron model, where all circuit elements (switch, resistors, and capacitors) are manufactured using vanadium oxides of various stoichiometry (VO, VO_2, V_2O_3, and V_2O_5).

All the results presented in the current study were obtained by modeling the VO_2 neuron using numerical methods in the LTspice simulator. However, the I–V characteristic of the VO_2 switch corresponding to the experimental data was used in the model (see Section 2.1). Therefore, a discussion of the physical mechanisms that affect the I–V characteristics is of great importance for predicting the areas of practical application and comparing the characteristics of neuron models.

An increase in the ambient temperature T_0 leads to a decrease in the threshold voltage V_{th}, and at $T_0 \sim T_{th}$, the effect of electrical switching will be suppressed, because the VO_2 channel will achieve a highly conductive state [41,52,53]. It imposes a limitation on the use of a VO_2 neuron, where the operability will be limited at $T_0 > T_{th}$. The value of T_{th} is not high, and reaches $T_{th} \sim 340$ K (67 °C) and creates the question - how to increase T_{th}? A good overview of the T_{th} modulation methods by doping with various elements is presented in Reference [54]. For example, Cr doping can increase T_{th} by 10 °C [55]. An alternative way to increase T_{th} is to use other materials with an S-type I–V characteristic. NbO_2–based structures, having $T_{th} \sim 1070$ K, demonstrate electrical switching up to temperatures of $T_0 \sim 300$ °C [56], and can be used in the presented neuron model. The model is invariant to the use of other materials with the effect of electrical switching, and the main requirement is the presence of an S-type I–V characteristic. VO_2-based structures are a good model object and are often used in neural circuits. Nevertheless, the task of finding switching structures with a wide temperature range is a promising endeavour.

Another problem is the variation of V_{th} with the temperature. Therefore, if it is necessary to stabilize the operation of a neuron, it is necessary to come up with additional thermal compensation schemes.

To optimize the circuit presented in Figure 2, we propose to exclude the capacitor C_{sum} from the circuit. When switching a VO_2 switch by rectangular pulses, there is an effect of a time delay of switching on the switch. An inverse dependence of the time delay on the pulse amplitude is associated with the thermal heating of the switching channel to the phase transition temperature of the metal insulator. The physics of the electrical switching process is described in detail in Reference [27]. If pulse durations and coding time intervals Δt_{max} are used within the delay times of the switching, the pulse integration effect can be implemented without C_{sum} capacity, and will be caused by the heat accumulation in the region of the switching channel. A similar idea to use heat storage in the switch region to accumulate action in a neuron was proposed in Reference [31]. The role of capacitors in the oscillator circuit is discussed in a number of sources [25,57,58], and the oscillations can be obtained without an integrating capacitor, while being only due to the effects of heat storage. The study of the effect of temperature integration of input pulses could be the subject of future research.

The coupling between the switches in the network can be implemented not only by electrical coupling through resistors $R_{w\_i,j}$, but, as described in our previous studies, can be organized through the thermal coupling of the switches [59,60]. The development of spike neural networks with thermal coupling could be the subject of further research.

The reduction of SNN classification uncertainty, when applying a highly distorted template (the right image in Figure 11), by reducing the time interval for coding spikes Δt_{max}, has its own limitations. The current SNN model does not take into account the effect of the turn-on and turn-off delay of the switch described above. For example, the turn-on time of the VO_2 switch, using our input signal amplitudes, does not exceed 10 ns [25], while the turn-off time can be much longer (hundreds of nanoseconds). In the time scale of the current SNN model, operating in microsecond intervals, by taking into account the effect of the turn-on delay, does not affect the results of the SNN operation. However, the turn-on and turn-off times can vary significantly, when using other switches, resistors, and resistances. It should be taken into account when designing SNN.

An important characteristic of the network is the pattern recognition time [61]. The SNN architecture, which we propose, provides recognition time of the coding interval order, corresponding to 2–3 μs. After the recognition is completed, the system requires ~ 7–8 μs to reach the initial state, caused by the recharging of the capacities. Therefore, the current SNN is able to recognize up to 10^5 images per second, and its performance can be increased by reducing the capacitance rating and scaling the VO_2 switches [25]. The implemented method of information coding allows the use of single spikes. It does not only reduce the power consumption compared to the networks using rate coding, but minimizes the time to perform one image recognition operation.

At the end of this section, we present a comparison table of neurons with other proposed neuron devices. The neurons in Table 2 are divided into two groups: neurons based on silicon (CMOS) technology and neurons based on VO_2 switches. The main parameters of the neurons are the size of the active element and the energy consumption. Although the silicon neurons have an advantage in these parameters, in our study, the spikes duration is of the least importance. Another advantage of the VO_2 neuron compared to the CMOS neuron is, apparently, the high noise level of the current channel that leads to the stochastic behavior of the neuron described in References [33,34]. This property allows the network to escape local minima and reach the global minimum of the error surface.

Table 2. Comparison of neurons with other proposed neuron devices.

Device	Neuron Type Material/Platform	Active Element Size (a) and Neuron Area (S_{neuron})	Spike Amplitude (V_{spike}), Peak Power (P_{max}), Duration, (Δt_{spike}) and Energy per Spike (E_{spike})	Integration and Threshold Mechanism, Threshold Voltage of the Active Element V_{th}	SNN with Object Recognition, Coding Mechanism
VO_2 (current study)	Leaky Integrate and Fire Vanadium Dioxide (VO_2)	$a \sim 3$ μm	$V_{spike} = 3.2$ V Δt_{spike}~500 ns P_{max}~37 mW E_{spike}~ 18 nJ	Capacitor charging, Switching effect when reaching V_{th}, $V_{th}(VO_2)$~5.6 V	Time to first spike
Oxide neuron [35]	Piecewise linear FitzHugh-Nagumo, FitzHugh–Rinzel Vanadium Dioxide (VO_2), Niobium oxide (NbO)	$a \sim 3$ μm	V_{spike}~3.5V Δt_{spike}~100 μs P_{max}~72 mW E_{spike}~ 7 μJ	Capacitor charging and energy of inductance magnetic field, switching effect when reaching V_{th}, $V_{th}(VO_2)$~ 5.6 V $V_{th}(NbO_2)$~ 0.9 V	-
Stochastic VO_2 neuron [33]	Integrate and fire Vanadium Dioxide (VO2)	$a \sim 100$ nm	V_{spike}~0.5 V Δt_{spike}~4 μs P_{max}~12 μW E_{spike}~50 pJ	Capacitor charging, switching effect when reaching V_{th}, $V_{th}(VO_2)$~ 1.7 V	Rate coding
CMOS neuron [62]	Leaky Integrate and fire CMOS	$a \sim 90$nm S_{neuron}= 442 μm^2	$V_{spike} = 0.6$ V Δt_{spike}~3 ms $E_{spike} = 0.4$ pJ	Capacitor charging. Reset using comparator, V_{th}~ 0.6 V	-
CMOS neuron [63]	Simplified Morris - Lecar model CMOS	$a \sim 65$ nm $S_{neuron} = 35$ μm^2	$V_{spike} = 112$ mV Δt_{spike}~18 μs $E_{spike} = 4$ fJ	Capacitor charging and discharging through transistors, V_{th}~ 112mV	-

The spike amplitude, power, and energy consumption of the VO_2 neuron depend on the threshold switching voltage V_{th}, if only the energy release on the VO_2 switch is taken into account. The main technological parameters affecting the V_{th} value are the resistivity in the insulator phase ρ_{off} and the contact geometry [25,32]. In Reference [25], we obtained an equation for approximating V_{th}.

$$V_{th} = \frac{\sqrt{\lambda \cdot \rho_{off} \cdot (T_{th} - T_0)}}{\sqrt{d}} \cdot a^{\beta} \quad (1)$$

where d is the thickness of the VO_2 film, a is the inter-electrode distance, λ is the heat-transfer coefficient, and β is the exponential coefficient that determines the effective area of the heated zone in the inter-electrode gap ($\beta < 1$).

By decreasing the value of a, the structures with reduced V_{th} can be obtained. Using Equation (1) and the current value when the structure is turned on, estimated as $I_{on} = V_{th}/R_{on}$ (where $R_{on} = \rho_{on}/d$), we can propose an equation for the maximum power per spike ($\lambda = 35$ W/m · K, $\rho_{off} = 4 \cdot 10^{-2}$ Ω · m, $\rho_{on} = 4 \cdot 10^{-4}$ Ω · m, $\beta = 0.56$, $T_{th} = 340$ K, $T_0 = 300$ K [25]):

$$P_{max} = V_{th} \cdot I_{on} = \lambda \cdot \frac{\rho_{off}}{\rho_{on}} \cdot (T_{th} - T_0) \cdot a^{2\beta} \quad (2)$$

By reducing the size of the inter-electrode distance a and the ratio (ρ_{off}/ρ_{on}), we would significantly reduce the value of P_{max}. For example, at $a = 1$ μm, the maximum power P_{max} is ~ 26 mW, while at $a = 100$ nm, the maximum power P_{max} drops to 2 mW.

E_{spike} can be estimated by multiplying the spike duration by the maximum power, E_{spike}~ $P_{max} \cdot \Delta t_{spike}$. In Reference [25], we modelled the switch on and switch off durations of switches, which determine the minimum Δt_{spike} values. We demonstrated that the durations decrease with decreasing a. Therefore, it is possible to predict a significant decrease in E_{spike} with a decrease in the size of switching elements. The estimates are the following: $E_{spike} \sim 6.4$ nJ at $a = 1$ μm ($\Delta t_{spike} = 240$ ns), and $E_{spike} \sim 105$ pJ at $a = 100$ nm ($\Delta t_{spike} = 52$ ns).

The last column in Table 2 demonstrates that the majority of the previous studies gives only the model of VO_2 neuron itself. In the current study, we present the simple neural network that is capable of pattern recognition, using the timing method of information coding, which has a clear energy advantage over the firing rates coding method [33].

5. Conclusions

In the current study, we present the new model of an LIF neuron based on one switching VO_2 element. The neuron circuit was modeled in the LTspice program, and, for the component emulating the switch, a voltage-controlled key was used, which the I – V characteristic corresponded to experimental data. During the simulation, the VO_2-neuron model demonstrates biosimilar properties, such as spike latency, subthreshold oscillations, refractory period, threshold behavior, and spike frequency adaptation. A two-layer SNN was designed to allow pattern recognition. The coupling between the neurons of the input and output layers was implemented using excitatory connections, and, inside the output layer, the coupling used inhibitory connections. This architecture led to the activation of only one output neuron associated with the most similar pattern, according to the WTA principle. As an example, we studied the network that had nine input and threeoutput neurons, which was trained to recognize three patterns (3 × 3 pixels). A timing method of information coding was used, where the color intensity of the pixel was determined by the time delay between the spikes. The training was conducted using the supervised SPDT method, taking into account the time delay of pre-synaptic and post-synaptic spikes. To analyze the operation of the trained network, the images of distorted patterns from the training set were sent to the network input, and the images were correctly recognized in most cases. The network is capable of recognizing up to 10^5 images per second, and the classification process is highly dependent on the time parameters of the network and the effect of electrical switching. Network architecture has the potential for further scaling, which increases the speed of recognition and miniaturization of the components. In the future, we plan to continue the work toward optimization of both the neuron circuit and the network architecture for classifying images from standardized databases [64].

Author Contributions: Conceptualization, M.B. Methodology, M.B. and A.V. Software validation, M.B. Writing—original draft preparation, M.B. and A.V. Project administration, A.V.

Funding: The Russian Science Foundation (grant no. 16-19-00135) supported this research.

Acknowledgments: The authors express their gratitude to Andrei Rikkiev for the valuable comments in the course of the article translation and revision.

Conflicts of Interest: The authors declare no conflict of interest.

References

1. Min, E.; Guo, X.; Liu, Q.; Zhang, G.; Cui, J.; Long, J. A Survey of Clustering with Deep Learning: From the Perspective of Network Architecture. *IEEE Access* **2018**, *6*, 39501–39514. [CrossRef]
2. Cireşan, D.; Meier, U.; Schmidhuber, J. Multi-column deep neural networks for image classification. In Proceedings of the 2012 IEEE Conference on Computer Vision and Pattern Recognition, Providence, RI, USA, 16–21 June 2012; pp. 3642–3649.
3. Abdel-Hamid, O.; Mohamed, A.-R.; Jiang, H.; Deng, L.; Penn, G.; Yu, D. Convolutional Neural Networks for Speech Recognition. *IEEE/ACM Trans. Audio Speech Lang. Process.* **2014**, *22*, 1533–1545. [CrossRef]
4. Du, Y.; Wang, W.; Wang, L. Hierarchical recurrent neural network for skeleton based action recognition. In Proceedings of the 2015 IEEE Conference on Computer Vision and Pattern Recognition (CVPR), Boston, MA, USA, 7–12 June 2015; pp. 1110–1118.

5. Patterson, J.; Gibson, A. *Deep Learning: A Practitioner's Approach*; O'Reilly: Sebastopol, CA, USA, 2017; ISBN 9781491924570.
6. Liu, W.; Wang, Z.; Liu, X.; Zeng, N.; Liu, Y.; Alsaadi, F.E. A survey of deep neural network architectures and their applications. *Neurocomputing* **2017**, *234*, 11–26. [CrossRef]
7. Maass, W. Networks of spiking neurons: The third generation of neural network models. *Neural Netw.* **1997**, *10*, 1659–1671. [CrossRef]
8. Ghosh-Dastidar, S.; Adeli, H. spiking neural networks. *Int. J. Neural Syst.* **2009**, *19*, 295–308. [CrossRef] [PubMed]
9. Bonabi, S.Y.; Asgharian, H.; Safari, S.; Ahmadabadi, M.N. FPGA implementation of a biological neural network based on the Hodgkin-Huxley neuron model. *Front. Mol. Neurosci.* **2014**, *8*, 379.
10. Cheung, K.; Schultz, S.R.; Luk, W. *A Large-Scale Spiking Neural Network Accelerator for FPGA Systems*; Springer Science and Business Media LLC: Berlin, Germany, 2012; Volume 7552, pp. 113–120.
11. Beyeler, M.; Dutt, N.D.; Krichmar, J.L. Categorization and decision-making in a neurobiologically plausible spiking network using a STDP-like learning rule. *Neural Netw.* **2013**, *48*, 109–124. [CrossRef] [PubMed]
12. Diehl, P.U.; Cook, M. Unsupervised learning of digit recognition using spike-timing-dependent plasticity. *Front. Comput. Neurosci.* **2015**, *9*, 99. [CrossRef]
13. Beyeler, M.; Oros, N.; Dutt, N.; Krichmar, J.L. A GPU-accelerated cortical neural network model for visually guided robot navigation. *Neural Netw.* **2015**, *72*, 75–87. [CrossRef]
14. Tavanaei, A.; Ghodrati, M.; Kheradpisheh, S.R.; Masquelier, T.; Maida, A. Deep learning in spiking neural networks. *Neural Netw.* **2019**, *111*, 47–63. [CrossRef]
15. Jeong, D.S.; Kim, K.M.; Kim, S.; Choi, B.J.; Hwang, C.S. Memristors for Energy-Efficient New Computing Paradigms. *Adv. Electron. Mater.* **2016**, *2*, 1600090. [CrossRef]
16. Jeong, H.; Shi, L. Memristor devices for neural networks. *J. Phys. D Appl. Phys.* **2019**, *52*, 023003. [CrossRef]
17. Srinivasan, G.; Sengupta, A.; Roy, K. Magnetic Tunnel Junction Based Long-Term Short-Term Stochastic Synapse for a Spiking Neural Network with On-Chip STDP Learning. *Sci. Rep.* **2016**, *6*, 29545. [CrossRef] [PubMed]
18. Kim, H.; Hwang, S.; Park, J.; Park, B.-G. Silicon synaptic transistor for hardware-based spiking neural network and neuromorphic system. *Nanotechnology* **2017**, *28*, 405202. [CrossRef] [PubMed]
19. Wang, Z.; Joshi, S.; Savel'ev, S.; Song, W.; Midya, R.; Li, Y.; Rao, M.; Yan, P.; Asapu, S.; Zhuo, Y.; et al. Fully memristive neural networks for pattern classification with unsupervised learning. *Nat. Electron.* **2018**, *1*, 137. [CrossRef]
20. Wang, Z.; Crafton, B.; Gomez, J.; Xu, R.; Luo, A.; Krivokapic, Z.; Martin, L.; Datta, S.; Raychowdhury, A.; Khan, A.I. Experimental Demonstration of Ferroelectric Spiking Neurons for Unsupervised Clustering. In Proceedings of the 2018 IEEE International Electron Devices Meeting (IEDM), San Francisco, CA, USA, 1–5 December 2018; pp. 13.3.1–13.3.4.
21. Zhou, E.; Fang, L.; Yang, B. Memristive Spiking Neural Networks Trained with Unsupervised STDP. *Electronics* **2018**, *7*, 396. [CrossRef]
22. Jerry, M.; Tsai, W.-Y.; Xie, B.; Li, X.; Narayanan, V.; Raychowdhury, A.; Datta, S. Phase transition oxide neuron for spiking neural networks. In Proceedings of the 2016 74th Annual Device Research Conference (DRC), Newark, DE, USA, 19–22 June 2016; pp. 1–2.
23. Jeong, D.S.; Thomas, R.; Katiyar, R.S.; Scott, J.F.; Kohlstedt, H.; Petraru, A.; Hwang, C.S. Emerging memories: Resistive switching mechanisms and current status. *Rep. Prog. Phys.* **2012**, *75*, 76502. [CrossRef]
24. Li, Y.; Wang, Z.; Midya, R.M.; Xia, Q.; Yang, J.J. Review of memristor devices in neuromorphic computing: Materials sciences and device challenges. *J. Phys. D Appl. Phys.* **2018**, *51*, 503002. [CrossRef]
25. Pergament, A.; Velichko, A.; Belyaev, M.; Putrolaynen, V. Electrical switching and oscillations in vanadium dioxide. *Phys. B Condens. Matter* **2018**, *536*, 239–248. [CrossRef]
26. Crunteanu, A.; Givernaud, J.; Leroy, J.; Mardivirin, D.; Champeaux, C.; Orlianges, J.-C.; Catherinot, A.; Blondy, P. Voltage- and current-activated metal–insulator transition in VO_2 -based electrical switches: A lifetime operation analysis. *Sci. Technol. Adv. Mater.* **2010**, *11*, 065002. [CrossRef]
27. Belyaev, M.A.; Boriskov, P.P.; Velichko, A.A.; Pergament, A.L.; Putrolainen, V.V.; Ryabokon', D.V.; Stefanovich, G.B.; Sysun, V.I.; Khanin, S.D. Switching Channel Development Dynamics in Planar Structures on the Basis of Vanadium Dioxide. *Phys. Solid State* **2018**, *60*, 447–456. [CrossRef]

28. Yi, W.; Tsang, K.K.; Lam, S.K.; Bai, X.; Crowell, J.A.; Flores, E.A. Biological plausibility and stochasticity in scalable VO2 active memristor neurons. *Nat. Commun.* **2018**, *9*, 4661. [CrossRef] [PubMed]
29. Ignatov, M.; Ziegler, M.; Hansen, M.; Petraru, A.; Kohlstedt, H. A memristive spiking neuron with firing rate coding. *Front. Mol. Neurosci.* **2015**, *9*, 49. [CrossRef] [PubMed]
30. Lin, J.; Guha, S.; Ramanathan, S. Vanadium Dioxide Circuits Emulate Neurological Disorders. *Front. Mol. Neurosci.* **2018**, *12*, 856. [CrossRef] [PubMed]
31. Amer, S.; Hasan, M.S.; Adnan, M.M.; Rose, G.S. SPICE Modeling of Insulator Metal Transition: Model of the Critical Temperature. *IEEE J. Electron Devices Soc.* **2019**, *7*, 18–25. [CrossRef]
32. Lin, J.; Sonde, S.; Chen, C.; Stan, L.; Achari, K.V.L.V.; Ramanathan, S.; Guha, S. Low-voltage artificial neuron using feedback engineered insulator-to-metal-transition devices. In Proceedings of the 2016 IEEE International Electron Devices Meeting (IEDM), San Francisco, CA, USA, 3–7 December 2016; p. 35.
33. Jerry, M.; Parihar, A.; Grisafe, B.; Raychowdhury, A.; Datta, S. Ultra-low power probabilistic IMT neurons for stochastic sampling machines. In Proceedings of the 2017 Symposium on VLSI Technology, Kyoto, Japan, 5–8 June 2017; pp. T186–T187.
34. Parihar, A.; Jerry, M.; Datta, S.; Raychowdhury, A. Stochastic IMT (Insulator-Metal-Transition) Neurons: An Interplay of Thermal and Threshold Noise at Bifurcation. *Front. Neurosci.* **2018**, *12*, 210. [CrossRef] [PubMed]
35. Boriskov, P.; Velichko, A. Switch Elements with S-Shaped Current-Voltage Characteristic in Models of Neural Oscillators. *Electronics* **2019**, *8*, 922. [CrossRef]
36. Oster, M.; Douglas, R.; Liu, S.-C. Computation with Spikes in a Winner-Take-All Network. *Neural Comput.* **2009**, *21*, 2437–2465. [CrossRef]
37. Ponulak, F.; Kasinski, A. Introduction to spiking neural networks: Information processing, learning and applications. *Acta Neurobiol. Exp.* **2011**, *71*, 409–433.
38. Serrano-Gotarredona, T.; Masquelier, T.; Prodromakis, T.; Indiveri, G.; Linares-Barranco, B. STDP and STDP variations with memristors for spiking neuromorphic learning systems. *Front. Mol. Neurosci.* **2013**, *7*, 2. [CrossRef]
39. Izhikevich, E. Which Model to Use for Cortical Spiking Neurons? *IEEE Trans. Neural Netw.* **2004**, *15*, 1063–1070. [CrossRef] [PubMed]
40. Karda, K.; Mouli, C.; Ramanathan, S.; Alam, M.A. A Self-Consistent, Semiclassical Electrothermal Modeling Framework for Mott Devices. *IEEE Trans. Electron Devices* **2018**, *65*, 1672–1678. [CrossRef]
41. Pergament, A.; Boriskov, P.; Velichko, A.; Kuldin, N.; Velichko, A. Switching effect and the metal–insulator transition in electric field. *J. Phys. Chem. Solids* **2010**, *71*, 874–879. [CrossRef]
42. Querlioz, D.; Bichler, O.; Dollfus, P.; Gamrat, C. Immunity to Device Variations in a Spiking Neural Network with Memristive Nanodevices. *IEEE Trans. Nanotechnol.* **2013**, *12*, 288–295. [CrossRef]
43. Shukla, A.; Kumar, V.; Ganguly, U. A software-equivalent SNN hardware using RRAM-array for asynchronous real-time learning. In Proceedings of the 2017 International Joint Conference on Neural Networks (IJCNN), Anchorage, AK, USA, 14–19 May 2017; pp. 4657–4664.
44. Kwon, M.-W.; Baek, M.-H.; Hwang, S.; Kim, S.; Park, B.-G. Spiking Neural Networks with Unsupervised Learning Based on STDP Using Resistive Synaptic Devices and Analog CMOS Neuron Circuit. *J. Nanosci. Nanotechnol.* **2018**, *18*, 6588–6592. [CrossRef] [PubMed]
45. Yousefzadeh, A.; Stromatias, E.; Soto, M.; Serrano-Gotarredona, T.; Linares-Barranco, B. On Practical Issues for Stochastic STDP Hardware With 1-bit Synaptic Weights. *Front. Mol. Neurosci.* **2018**, *12*, 665. [CrossRef] [PubMed]
46. Saunders, D.J.; Siegelmann, H.T.; Kozma, R.; Ruszinkao, M. STDP Learning of Image Patches with Convolutional Spiking Neural Networks. In Proceedings of the 2018 International Joint Conference on Neural Networks (IJCNN), Rio de Janeiro, Brazil, 8–13 July 2018; pp. 1–7.
47. Bichler, O.; Querlioz, D.; Thorpe, S.J.; Bourgoin, J.-P.; Gamrat, C. Extraction of temporally correlated features from dynamic vision sensors with spike-timing-dependent plasticity. *Neural Netw.* **2012**, *32*, 339–348. [CrossRef] [PubMed]
48. Tavanaei, A.; Maida, A. BP-STDP: Approximating backpropagation using spike timing dependent plasticity. *Neurocomputing* **2019**, *330*, 39–47. [CrossRef]
49. Nishitani, Y.; Kaneko, Y.; Ueda, M. Supervised Learning Using Spike-Timing-Dependent Plasticity of Memristive Synapses. *IEEE Trans. Neural Netw. Learn. Syst.* **2015**, *26*, 2999–3008. [CrossRef] [PubMed]

50. Lee, C.; Panda, P.; Srinivasan, G.; Roy, K. Training Deep Spiking Convolutional Neural Networks With STDP-Based Unsupervised Pre-training Followed by Supervised Fine-Tuning. *Front. Mol. Neurosci.* **2018**, *12*, 435. [CrossRef]
51. Garcia, C.; Delakis, M. Convolutional face finder: A neural architecture for fast and robust face detection. *IEEE Trans. Pattern Anal. Mach. Intell.* **2004**, *26*, 1408–1423. [CrossRef] [PubMed]
52. Kim, S.; Lin, C.-Y.; Kim, M.-H.; Kim, T.-H.; Kim, H.; Chen, Y.-C.; Chang, Y.-F.; Park, B.-G. Dual Functions of V/SiOx/AlOy/p++Si Device as Selector and Memory. *Nanoscale Res. Lett.* **2018**, *13*, 252. [CrossRef] [PubMed]
53. Lin, C.-Y.; Chen, P.-H.; Chang, T.-C.; Chang, K.-C.; Zhang, S.-D.; Tsai, T.-M.; Pan, C.-H.; Chen, M.-C.; Su, Y.-T.; Tseng, Y.-T.; et al. Attaining resistive switching characteristics and selector properties by varying forming polarities in a single HfO2-based RRAM device with a vanadium electrode. *Nanoscale* **2017**, *9*, 8586–8590. [CrossRef] [PubMed]
54. Sun, C.; Yan, L.; Yue, B.; Liu, H.; Gao, Y. The modulation of metal–insulator transition temperature of vanadium dioxide: A density functional theory study. *J. Mater. Chem. C* **2014**, *2*, 9283–9293. [CrossRef]
55. Brown, B.L.; Nordquist, C.D.; Jordan, T.S.; Wolfley, S.L.; Leonhardt, D.; Edney, C.; Custer, J.A.; Lee, M.; Clem, P.G. Electrical and optical characterization of the metal-insulator transition temperature in Cr-doped VO2 thin films. *J. Appl. Phys.* **2013**, *113*, 173704. [CrossRef]
56. Pergament, A.; Stefanovich, G.; Malinenko, V.; Velichko, A. Electrical Switching in Thin Film Structures Based on Transition Metal Oxides. *Adv. Condens. Matter Phys.* **2015**, *2015*, 654840. [CrossRef]
57. Lepage, D.; Chaker, M. Thermodynamics of self-oscillations in VO_2 for spiking solid-state neurons. *AIP Adv.* **2017**, *7*, 055203. [CrossRef]
58. Sakai, J. High-efficiency voltage oscillation in VO_2 planer-type junctions with infinite negative differential resistance. *J. Appl. Phys.* **2008**, *103*, 103708. [CrossRef]
59. Velichko, A.; Belyaev, M.; Putrolaynen, V.; Perminov, V.; Pergament, A. Thermal coupling and effect of subharmonic synchronization in a system of two VO_2 based oscillators. *Solid-State Electron.* **2018**, *141*, 40–49. [CrossRef]
60. Velichko, A.; Belyaev, M.; Putrolaynen, V.; Perminov, V.; Pergament, A. Modeling of thermal coupling in VO_2 -based oscillatory neural networks. *Solid-State Electron.* **2018**, *139*, 8–14. [CrossRef]
61. Savchenko, A. Sequential three-way decisions in multi-category image recognition with deep features based on distance factor. *Inf. Sci.* **2019**, *489*, 18–36. [CrossRef]
62. Cruz-Albrecht, J.M.; Yung, M.W.; Srinivasa, N. Energy-Efficient Neuron, Synapse and STDP Integrated Circuits. *IEEE Trans. Biomed. Circuits Syst.* **2012**, *6*, 246–256. [CrossRef] [PubMed]
63. Sourikopoulos, I.; Hedayat, S.; Loyez, C.; Danneville, F.; Hoel, V.; Mercier, E.; Cappy, A. A 4-fJ/Spike Artificial Neuron in 65 nm CMOS Technology. *Front. Mol. Neurosci.* **2017**, *11*, 1597. [CrossRef] [PubMed]
64. LeCun, Y.; Cortes, C.; Burges, C. MNIST Handwritten Digit Database. Available online: http://yann.lecun.com/exdb/mnist/ (accessed on 9 November 2018).

Article

Benchmarking a Many-Core Neuromorphic Platform With an MPI-Based DNA Sequence Matching Algorithm

Gianvito Urgese [1,*], Francesco Barchi [2], Emanuele Parisi [2], Evelina Forno [2], Andrea Acquaviva [3] and Enrico Macii [1]

1 Interuniversity Department of Regional and Urban Studies and Planning, Politecnico di Torino, 10129 Torino, Italy; enrico.macii@polito.it
2 Department of Control and Computer Engineering, Politecnico di Torino, 10129 Torino, Italy; francesco.barchi@polito.it (F.B.); emanuele.parisi@polito.it (E.P.); evelina.forno@polito.it (E.F.)
3 Department of Electrical, Electronic, and Information Engineering "Guglielmo Marconi", University of Bologna, 40126 Bologna, Italy; andrea.acquaviva@unibo.it
* Correspondence: gianvito.urgese@polito.it

Received: 29 September 2019; Accepted: 12 November 2019; Published: 14 November 2019

Abstract: SpiNNaker is a neuromorphic globally asynchronous locally synchronous (GALS) multi-core architecture designed for simulating a spiking neural network (SNN) in real-time. Several studies have shown that neuromorphic platforms allow flexible and efficient simulations of SNN by exploiting the efficient communication infrastructure optimised for transmitting small packets across the many cores of the platform. However, the effectiveness of neuromorphic platforms in executing massively parallel general-purpose algorithms, while promising, is still to be explored. In this paper, we present an implementation of a parallel DNA sequence matching algorithm implemented by using the MPI programming paradigm ported to the SpiNNaker platform. In our implementation, all cores available in the board are configured for executing in parallel an optimised version of the *Boyer-Moore* (BM) algorithm. Exploiting this application, we benchmarked the SpiNNaker platform in terms of scalability and synchronisation latency. Experimental results indicate that the SpiNNaker parallel architecture allows a linear performance increase with the number of used cores and shows better scalability compared to a general-purpose multi-core computing platform.

Keywords: benchmarking neuromorphic HW; neuromorphic platform; spiNNaker; spinMPI; MPI for neuromorphic HW; Boyer-Moore; DNA matching algorithm

---

1. Introduction

A neuromorphic system is a massively multi-core system composed of simple processing units and memory elements communicating by message exchanging [1]. This type of approach strives to simulate the behaviour of the brain using design principles based on biological nervous systems. Neuromorphic systems differ from traditional multi-core systems in the way in which memory and processing are organised. Indeed, in this case, memory is distributed with processing units rather than centralised and physically separated from the cores. Using this strategy, it is possible to avoid the traditional bottleneck of memory access time, present in the classical Von-Neumann architectures. The main idea behind this kind of system is to process information using an event-driven protocol that lets the cores work in an asynchronous way [2]. The processing units remain in an idle state until an event is presented, triggering a reaction; after that, the units return to the idle state. Using this feature, neuromorphic systems are much more energy-efficient than traditional multi-core systems. This idea is inspired by biology; indeed, the human brain is composed of billions of neurons connected by synapses, working asynchronously, with a power consumption lower than that of a light-bulb [3]. Another peculiarity of neuromorphic systems is the high number of interconnections between the processing units, which speeds up and simplifies communication between the cores.

Neuromorphic HW platforms are attracting the interest of many research groups, mainly for the simulation of neural network structures observed in the brain and modelled through the simulation of Spiking Neural Networks (SNN). Although initially intended for brain simulations, the adoption of emerging neuromorphic HW architectures is also appealing in fields such as high-performance computing and robotics [4]. It has been proved that neuromorphic platforms provide better scalability than traditional multi-core architectures and are well suitable for classes of problems which require massive parallelism as well as the exchange of small messages, for which neuromorphic HW has a native optimised support [5]. However, the tools currently available in this field are still weak and miss many useful features required to support the spreading of a new neuromorphic-based computational paradigm.

In this paper, we analyse and benchmark the scaling capability of the SpiNNaker neuromorphic architecture. The SpiNNaker Machine is a multi-chip, globally asynchronous locally synchronous (GALS) neuromorphic architecture that connects general purpose ARM cores in a toroidal-shaped triangular mesh. It is efficient when used to solve problems modelled as a directed graph with an important communication component.

Other works have used this platform to execute parallel general purpose computation, with positive outcomes both for scaling performances and energy efficiency. In Blin et al. [5], authors have customised the neural model of an SNN configured for reproducing the connection graph of a page rank problem, showing that the scalability rate of the neuromorphic platform outperforms the general purpose architectures; whereas Sugiarto et al. [6] have implemented on SpiNNaker an energy efficient image processing algorithm, using a task graph representation to describe the mechanism and behavior of the method. However, none of these two approaches has tested synchronous applications, since both of them used an adapted SNN simulated with the standard asynchronous framework.

In previous work [7], authors have used a minimal Message Passing Interface (MPI) framework to implement a synchronization strategy that allows configuration of the cores of the board with a distributed application implementing the N-Body problem. The authors benchmarked the performance of the board in the execution of an MPI parallel application that simulates 2 k particles on 240 processors with a speed-up of 194× and an efficiency of 80% when compared to the serial version running on a single CPU.

In this paper, we compared the scaling performance of the SpiNNaker system with that offered by a many-core general purpose architecture. We implemented a parallel processing approach for a pattern matching algorithm able to identify the similarity of DNA sequences. In our implementation, we used the Message Passing Interface (MPI), a distributed parallel programming paradigm, to synchronise the communication of the computing cores on the two architectures. By using the MPI framework, we can port on the SpiNNaker platform an algorithm normally executed on a standard architecture without any need to re-shape the algorithm in the form of a Spiking Neural Network. The focus of the research presented in this paper is threefold.

- To benchmark the performances of the SpiNNaker board in computing pattern matching tasks by running synchronous data-stream algorithms.
- To explore the potential of the custom shape mesh, implemented on the SpiNNaker board, in a supporting parallel application that adopts a one-to-many communication system.
- To demonstrate how it is easy to port synchronous applications, implemented for the general-purpose computer, on the SpiNNaker board by using our software component that supports MPI for SpiNNaker.

The rest of the manuscript is organized as follows: Section 2 provides background information on existing neuromorphic architectures, with a detailed focus on the SpiNNaker board and on the DNA search algorithm. Section 3 describes the materials and methods used to carry out the study, whereas Section 4 examines experimental results. Finally, Section 5 closes with the conclusions.

2. Background

In the following, we provide a background on neuromorphic hardware in general and SpiNNaker in particular. Then we discuss the variant of the Boyer-Moore algorithm that we implemented with the MPI framework in order to benchmark the scaling capability of the SpiNNaker platform.

There are two main approaches to neuromorphic computing—VLSI architectures where neurons are modelled at transistor-level and communications are handled with connection crossbar array and custom architectures where general-purpose cores are connected to form a mesh of processors optimised for the transmission of small packets [8–10]. In the following, we report four representative architectures.

BrainScaleS is a VLSI platform developed at the University of Heidelberg [11]. The main idea behind this project is to use above-threshold analogue circuits to physically model neuronal processes, exploiting analogy between electronic circuits and the ionic circuits in biological neurons. Analogue neurons are delivered using wafer-scale integration.

Dynap-SEL is a VLSI chip called Dynamic Asynchronous Processor Scalable and Learning that is produced with four neural processing cores which implement 256 analog Adaptive Exponential Integrate and Fire neurons placed in a 16×16 grid with 64 programmable synapses for each neuron. In the Dynap-SEL architecture, it is available also a supplementary core 64 analog neurons and 8192 plastic synapses with on-chip learning and 4096 programmable synapses [12].

Loihi is a neuromorphic processor produced by Intel [13]. It features a many-core mesh comprising 128 neuromorphic cores, three embedded x 86 processor cores and off-chip communication interfaces that extend the mesh in 4-planar directions to other chips. All logic in the chip is digital and implemented as an asynchronous bundled-data design.

The Spiking Neural Network Architecture (*SpiNNaker*) [14] is a real-time neural network simulator following an event-driven computational approach [15]. This architecture is able to emulate neural populations and to simulate an entire Spiking Neural Network (SNN) in real-time. What sets SpiNNaker apart from all the above platforms is the fact that its architecture does not implement neurons via custom VLSI designed circuits, but it consists of a mesh of general-purpose ARM cores with a neuromorphic connectivity scheme. While the platform is designed to run SNN simulations and a software stack is provided to facilitate this purpose, in principle, the general-purpose cores can run any sort of C program compiled for ARM.

2.1. SpiNNaker Architecture

The base element of the SpiNNaker architecture is the SpiNNaker chip Figure 1, an SoC composed by 18 ARM-968 cores running at 200 MHz without a floating point unit but equipped with a custom router. Each processor has 32 kB of ITCM, 64 kB of DTCM, and shares through a system NoC 128 MB of SDRAM with the other processors in the chip. All the cores in the SoC (Application Processors) can run user applications, except one core for each chip, which is designated to be the *Monitor Processor*. This particular processor always executes the *SC&MP* program, which is a sort of operating system performing operations of memory management and acting as a packet manager, able to receive and transmit packet traffic from/to the cores. SpiNNaker chips (nodes) are connected to six neighbours and assembled on a PCB board made of 48 SpiNNaker chips (Spin5). The host computer can communicate with and configure a Spin5 via the Monitor Processor of the chip (0,0), the only one that is physically connected to an 100 Mbit Ethernet interface.

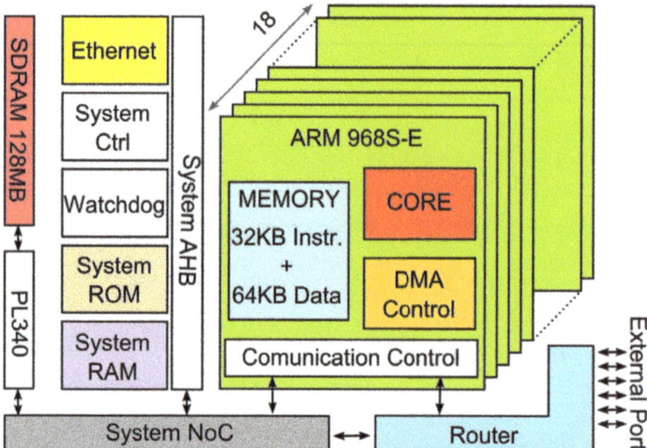

Figure 1. The SpiNNaker chip architecture.

2.2. SpiNNaker Network

The kernel of the interconnection among all cores of all chips of the simulator is the router, specifically designed to deliver packets as fast as possible (0.1 µs per hop) [16]. The particular design of the router, despite limitations on the synchronous transmission of packets [17,18], allows transmission of two operative packet types—Multicast (MC) and Point to Point (P2P). The length of these packets can be up to 72 bits and can carry a 32 bits long payload.

Multicast (MC) packets can reach many cores across the board. In neural simulations, they are widely used in order to spread neural potentials to multiple destinations. Point-to-Point (P2P) packets can be used for chip-to-chip transmissions. Each chip is uniquely identified by its coordinates (x, y), which define the chip's position in the chip mesh. P2P packets are always delivered to a chip's monitor processors.

The APIs of the SpiNNaker system provide a higher level of abstraction that simplifies the usage of chip interconnection. The SpiNNaker Datagram Protocol (SDP) can be used to manage communication between processors up to 256 Bytes [14]. The Monitor Processors act as a middleware between the SDP protocol and the on-board network. A Monitor Processor that receives an SDP packet splits the whole frame into 32-bit fragments to be delivered in the internal network through the P2P packets.

2.3. SpiNNaker Software

The software used to run a simulation managing the boards involves board-side code developed in C and Assembly [19] and host-side code mostly written in Python [20].

In this work we used the software stack provided by the SpinMPI library—a partial implementation of MPI on SpiNNaker [7] able to fully exploit the communication potential provided by the architecture, using the Application Command Framework (ACF) and the Multicast Communication Middleware (MCM) to manage communications.

The ACF uses the Application Command Protocol (ACP) to implement a Remote Procedure Call (RPC) capability in SpiNNaker at the application level [21]. Moreover, this library implements the *memory entity* concept. A memory entity is a managed memory space (DTCM, SysRAM, SDRAM), identified by an integer number, on which it is possible to perform CRUD operations (Create, Read, Update, Delete) locally or remotely. A *memory entity* can be created with a size limit of 256 Byte, that is, the ACP payload limit. The MCM instead implements unicast and broadcast communications, exploiting the multicast network capabilities of SpiNNaker.

2.4. The DNA Pattern Matching Algorithm

One of the most recurrent and widely studied problems in computer science is pattern matching—this problem has several real-world applications such as fast sub-string searching for network intrusion detection, mail spam filters, protein motif search and DNA/RNA sequence alignments [22]. Given a text string T of length n and a pattern string P of length $m \leq n$, the pattern matching problem can be stated as retrieving all positions i where pattern P occurs in text T, such that $0 \leq i \leq n - m$.

A straightforward solution for the pattern matching problem consists of looking for the pattern sequence in the text position by position until every occurrence is found. Unfortunately, such an approach leads to a $O(m \cdot n)$ asymptotic complexity, which is not acceptable for large sets of data.

Given the practical relevance of this problem, many approaches were proposed in the literature for improving the naïve way. One of these is the *Boyer-Moore* algorithm [23,24], which trades space usage for time efficiency, defining rules for pruning the search space avoiding the exploration of all text positions. A C++ implementation is available in Reference [25].

Figure 2 provides an intuition for this approach; given the text in the picture, the first attempt looks for pattern "GTA" in position ⓪, which is not correct.

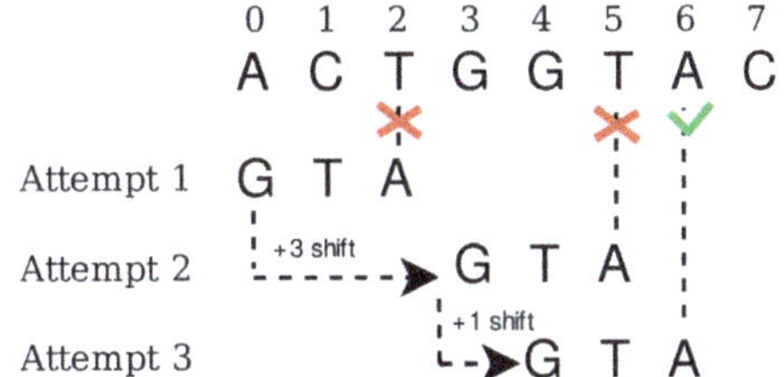

Figure 2. Intuition of the Boyer-Moore search procedure.

The naïve approach would perform the next search from position ①, but this is not ideal since the first instance of the letter "G" in the pattern occurs at position ③ in the text, meaning that searching any position in the middle is useless. Implementing this optimization requires pre-processing of the pattern to be matched; a *shift table* is computed, storing the number of text positions that can be safely skipped for each symbol in the target alphabet. Whenever a mismatch is found, given the next symbol to be searched, the *shift table* is accessed and the next position to be considered is computed.

We used a refined version of the *Boyer-Moore* algorithm, also known as *Fast string matching method for Encoded DNA sequences (FED)* [26], which takes advantage of the low-cardinality of the DNA alphabet. In the *FED* version, each of the four symbols composing the DNA alphabet is assigned a unique 2-bit code, packing four elements into a single byte, padding last bits with zeros in the case of sequences where the length is not a multiple of 4. Additionally, a bit-mask is used to distinguish valid bits from padding in the last encoded byte.

The procedure consists of two successive steps:

- *Pre-processing*, where texts and patterns are encoded and a *shift table* is computed for every pattern to be matched.
- *Matching*, where the actual search procedure is performed, is implemented as a byte-by-byte comparison between the text and pattern encoded sequences. If every byte of the pattern is sequentially found in the text, then the current position is registered as a match. Otherwise, the *shift table* is accessed to compute how many positions the pattern is allowed to skip before performing the next check.

Figure 3 summarizes the string matching procedure flow. As long as the customised *Boyer-Moore* procedure can perform a matching operation on encoded sequences, the encoding step can be considered not part of the algorithm as it can be done offline by storing the encoded sequences in custom binary files which constitute the actual source of data for the pattern matching engine.

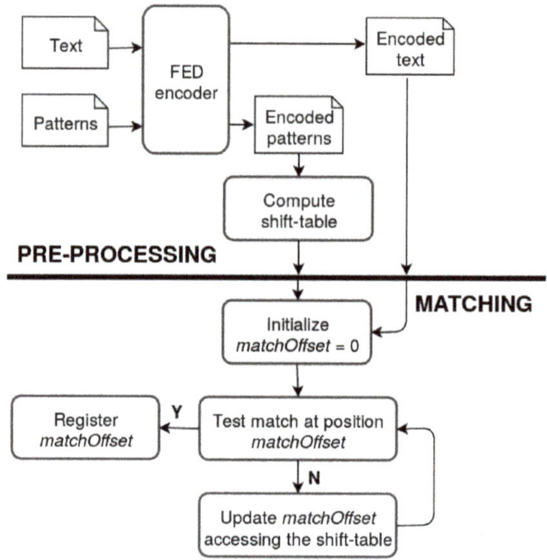

Figure 3. Flowchart of the string matching algorithm.

3. Materials and Methods

3.1. Implementation of DNA Sequence Matching with MPI

Pattern matching over DNA sequences can be considered an embarrassingly parallel application, because the average use case consists in matching millions of patterns against multiple text sequences, independently [27].

The inputs for the benchmark application are two binary files, storing the encoded texts and patterns to be analyzed. From an algorithmic point of view, running *FED* on already encoded sequences is equivalent to loading plain sequences and encoding them online. For the sake of benchmarking the communication effort in the target platforms, we decided to encode sequences off-line. Moreover, we split text sequences into a set of chunks with a given fixed size. This step is required because in bioinformatics applications, generally, the text represents one or more genomes and its size is not suitable to be sent in a single shot as it is.

Our parallel implementation of the search algorithm identifies two main roles among the MPI processes—the *MPI control process*, which is the role adopted by the MPI process with rank 0, and the *MPI worker*, which is the role adopted by all remaining MPI processes.

The algorithm works in two distinct steps, outlined in Figure 4: configuration Ⓐ and match Ⓑ. During configuration step Ⓐ, the *MPI control process* accesses the file system, loads the *FED* encoded patterns and distributes them among the *MPI workers* so that each working process handles approximately the same workload. Pattern distribution is implemented as a set of point-to-point communications, using MPI_Send/MPI_Recv primitives. Once an *MPI worker* receives its patterns it computes the *shift table* for them, completing the pre-processing phase shown in Figure 4. This strategy allows both to reduce the amount of data sent over the communication network, as the patterns are already encoded and to distribute the pre-processing efforts equally among all available working nodes, as long as any *MPI worker* finalizes the pre-processing step on its patterns only.

During the matching step (B), the *MPI control process* loads the encoded chunks of text and broadcasts them one at a time to all the *MPI workers*, which are in charge of performing the actual pattern matching procedure by calling the search primitives. As shown in Figure 4, once a match is found, it is saved into a buffer local to the MPI instance that discovered it. Once every chunk has been analysed, all the MPI instances synchronize to produce two report files containing information about the matches found and the run-time needed for accomplishing their tasks.

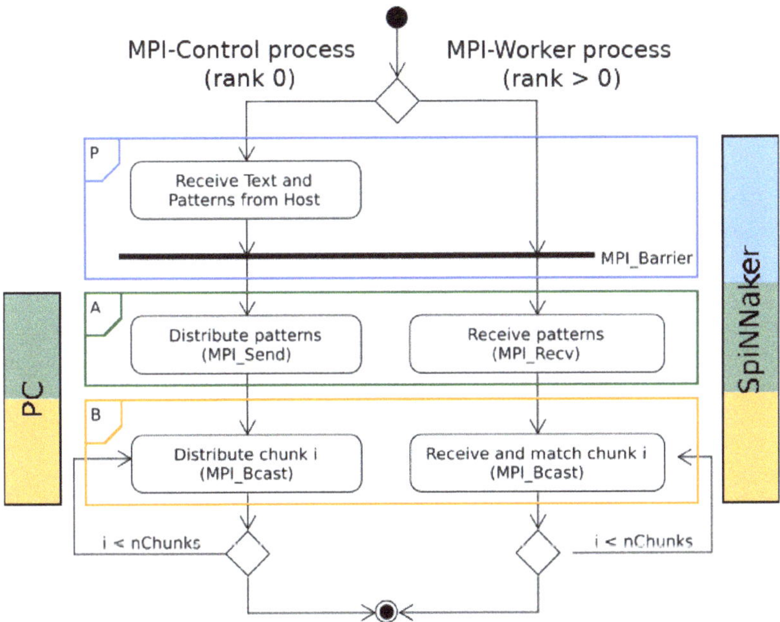

Figure 4. Flowchart of the implementation of MPI-FED on a general purpose architecture and on SpiNNaker. The step A performs the configuration, the step B execute the matching, whereas during the step P our implementation implement a preliminary phase for transferring the data to the SpiNNaker board.

3.2. *Adaptation of FED with MPI for SpiNNaker*

The implementation of *FED* with MPI for SpiNNaker retains the configuration (A) and match (B) phases from the previous section, as depicted in Figure 4. However, an additional preliminary phase (P) is required in order to transfer the problem data to the board. The configuration step (A) will then be performed by one of the SpiNNaker cores, taking up the role of *MPI control process*.

Using the SpinMPI Python library, the host launches the *MPI Runtime* and creates an *MPI Context* declaring the number of chips and cores that will be used by the application on the Spin5 board. The *MPI Runtime* is also in charge of loading and starting the application binary on the board.

In the preliminary phase (P), the communication between the computer host and the on-board application is performed through the use of ACP memory entities (MEs). First, the binary files containing the genome and the search patterns are read by the *MPI Runtime*. In this phase, the host will write into a ME belonging to processor (0, 0, 1) (the *MPI control process*) two integers indicating the number of chunks (`nchunks`) and patterns (`npatterns`) which will be loaded into SpiNNaker. The *MPI control process* allocates in SDRAM the memory necessary to contain all chunks and patterns. After allocation is performed, the addresses of these memory blocks are read by the *MPI Runtime*, again using ACP. The *MPI Runtime* can proceed to fill the *MPI control process* memory with the genome and the search patterns previously read.

An MPI Barrier forces all *MPI workers* to wait until the *MPI control process* has received all data from the *MPI Runtime*. Once the problem data has been transferred (phase Ⓟ), phase Ⓐ can begin. The *MPI control process* distributes the patterns among all worker cores through MPI_Send/MPI_Recv primitives and the *MPI workers* store the pattern data in their DTCM and compute the *shift tables*.

The phase Ⓑ begins after all patterns have been distributed. The *MPI control process* sends a text chunk to all *MPI workers* executing a broadcast communication. The SpiNNaker implementation of the *MPI_Bcast* function is a blocking call, as the memory limitations of the platform do not allow for large communication buffers; hence, the *MPI control process* will proceed to send the next chunk only after all workers have processed the current chunk. On the worker side, only one text buffer is allocated into DTCM, since the text chunks will be processed sequentially and a chunk can be replaced whenever a new one is obtained. When a *MPI worker* executing the *FED* algorithm finds a match position, it is stored into a linked list together with the chunk and matching pattern identifiers. Thus the position in the reference sequence can be retrieved.

After all the text chunks have been processed, the application is finalised, and the *MPI Runtime* can download the results directly from the memory of SpiNNaker cores.

4. Results and Discussion

In this section we report the results of tests designed to characterise the performance of the SpiNNaker system running a pattern matching algorithm implemented with the MPI programming paradigm.

As a preliminary evaluation, we measured the execution time and memory usage of the MPI primitives implemented on SpiNNaker that we will use for implementing the parallel FED algorithm. MPI_BARRIER, MPI_SEND/MPI_RECV, and MPI_BROADCAST primitives will be tested and the results are reported in Section 4.1. Next, we performed an evaluation of the performance of FED algorithm implemented with MPI and executed on the SpiNNaker and on the CPU-based architectures. This last analysis aims to evaluate the scalability and power efficiency of the SpiNNaker platform when compared with a standard architecture. The results are detailed in Section 4.2.

4.1. Performance of MPI on SpiNNaker

In Tables 1 and 2 and Figure 5 we report the performance of the MPI primitives on SpiNNaker. Table 1 shows the average execution time for 2000 iterations of the MPI_BARRIER synchronization primitive; the growth of the execution time is bounded with respect to the size of the context (i.e., the number of cores being used). The amount of memory necessary to store information about the context also grows slowly with respect to the number of cores.

Table 1. Table profiling the performance of 2000 iterations of MPI_BARRIER on SpiNNaker.

Cores	Time (ms)	Memory Usage
2	4.0	0.128
192	18.0	0.165
384	20.0	0.204
576	21.0	0.242
768	22.79	0.280

Table 2 shows the average execution time for the MPI_SEND/MPI_RECV unicast primitive. The average execution time grows linearly with the amount of data sent.

Finally, in Figure 5 we describe the average execution time for 2000 iterations of the MPI_BROADCAST primitive with respect to the amount of data sent and the context size. Once again the execution time grows linearly with the data sent, with overhead corresponding to the context-wide synchronization. The execution time also has a bounded growth in relation to the number of cores.

Table 2. Table profiling the performance of the MPI_SEND/MPI_RECV unicast primitive for different amounts of data sent on SpiNNaker.

Data Size	Time (ms)
1 kB	2.07
2 kB	4.12
4 kB	8.24

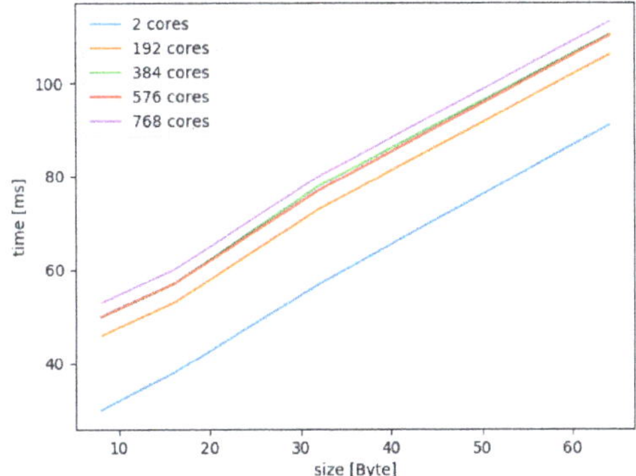

Figure 5. Graph profiling the performance of the MPI_BROADCAST primitive on SpiNNaker.

4.2. Evaluation of Boyer-Moore MPI Implementation Running on SpiNNaker

In the following, we analyse the efficiency and scalability of our optimised *Boyer-Moore* (*FED*) implementation on SpiNNaker. We compare it with the scalability on a traditional multi-core CPU using a server configuration with two Intel Silver Xeon 4114 processors, each with 10 cores and 20 threads. The *FED* algorithm is implemented in C and used to benchmark both Server and SpiNNaker architectures. The benchmark running on the general purpose Server architecture is written in C++ and compiled with g++ 7.4.0 and MPICH 3.3 parallel environment. The benchmark running on SpiNNaker architecture is written in C and compiled with gcc-arm-none-eabi 5.4.1 and SpinMPI 19w19. At this point, it is important to note that, by using the SpinMPI library, we ported the *FED* code written for a standard PC to the SpiNNaker hardware without applying any adaptation or transformation of the code.

The text used for the sake of testing is the *Escherichia coli* genome, which is about 4 million symbols long, leading to an encoded text of about 1 MB size, which is then split into a set of about 4000 chunks, each 256 Bytes long.

There exist two types of strategies to evaluate the scalability of a problem in a parallel environment:

- *Strong-scaling* [28] keeps the size of the problem fixed and evaluates the application runtime when multiple processes are used. This strategy is suitable for CPU-bounded problems.
- *Weak-scaling* [29] is used to test the scalability of memory-bounded problems, as it keeps constant the ratio between the problem size and the number of working processes used.

The SpiNNaker platform provides a fast, core-local data memory (DTCM) of 64 kB. This memory constraint allows to store at most 100 *FED* patterns per node, totalling 40 kB in size. Given this memory constraint, we decided to use a *weak-scaling* benchmarking strategy to scale our benchmark up to the 768 nodes available on SpiNNaker. The problem size must be calibrated in order to claim a condition of equivalence and perform a fair comparison between different architectures; in our case, a condition of equivalence is met whenever the same *FED* execution time t_{FED} is observed using a single *FED* worker. When SpinMPI is requested to match 1000 *FED* chunks against 100 *FED* patterns on a single node, a run-time of 26,970 ms is measured; the same run-time, for the MPICH implementation with 1000 *FED* chunks, is obtained when the single *FED* worker used is in charge of 12,500 *FED* patterns. This preliminary assessment is needed to evaluate only the scalability features of the two architectures, without considering the difference in computing power of the single working node for the two architectures. The reason for this comparison is to put the performance of MPI on SpiNNaker in a familiar perspective, as the CPU-DualSocket server is a widespread general purpose machine that allows to use MPI; however, the communication on the Xeon is networkless message passing happening entirely in RAM, while the message passing on SpiNNaker makes efficient use of the board's interconnection scheme.

A general strategy for evaluating the parallel scaling of an MPI application is computing the scaling efficiency, which measures how good the application is at using every node the parallel environment has. Given an environment with N workers and a problem that requires $t_{\text{FED},i}$ units of time to be solved with i workers, the *weak-scaling* efficiency E_N can be measured as in Equation (1). The speed-up S_N can be easily inferred from the efficiency and computed with Equation (2).

$$E_N = \frac{t_{\text{FED},1}}{t_{\text{FED},N}} \tag{1}$$

$$S_N = E_N \cdot N \tag{2}$$

Figures 6 and 7 report the speed-up and efficiency of the *FED* with MPI algorithm on the Server and SpiNNaker architectures. The horizontal axis represents the number of MPI workers used; both systems were tested until saturation, with the Server reaching 40 parallel workers through Intel hyper-threading and the Spin5 board utilizing all 768 available physical cores. Tests were performed for genomes of 500, 1000 and 2000 chunks.

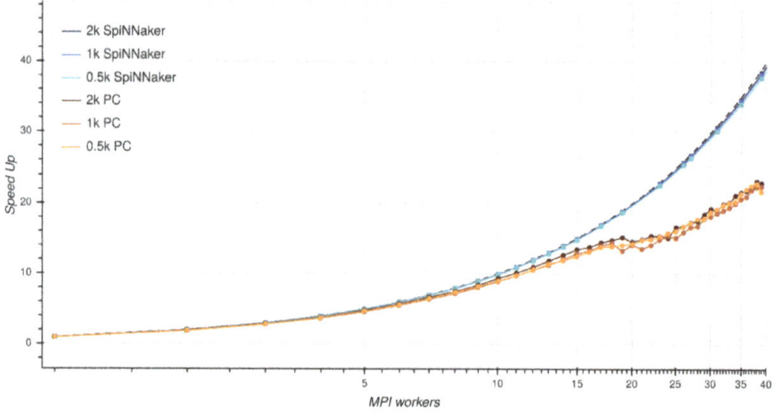

Figure 6. Comparison of Weak-scaling speed-up for MPI-FED on a general purpose architecture and on SpiNNaker.

In Figure 6 we can see how the massively parallel architecture of SpiNNaker influences the speed-up. The high number of physical cores on the machine lets the speed increase linearly, avoiding the discontinuities that a general-purpose processor has at critical points when hyperthreading is activated to provide the required number of workers (note, in the graph, the inflection point at 20 MPI workers for the PC version, i.e., the point at which the maximum number of physical threads on the Xeon is reached).

In Figure 7 SpiNNaker demonstrates excellent scalability, with efficiency values close to 95% for up to 200 workers. Additionally, we can see that the performance markedly improves for longer text sequences; the efficiency for 768 workers processing 2000 chunks is 87.83%. The reason for this happening is that as the size of the data to be processed increases, the ratio of processing time to communication time in the overall algorithm increases, since the data are only sent once at the beginning of processing and then gathered at the end. The bottleneck due to the communication overhead thus becomes less prevalent, and the efficiency improvement due to massive parallelism is more evident.

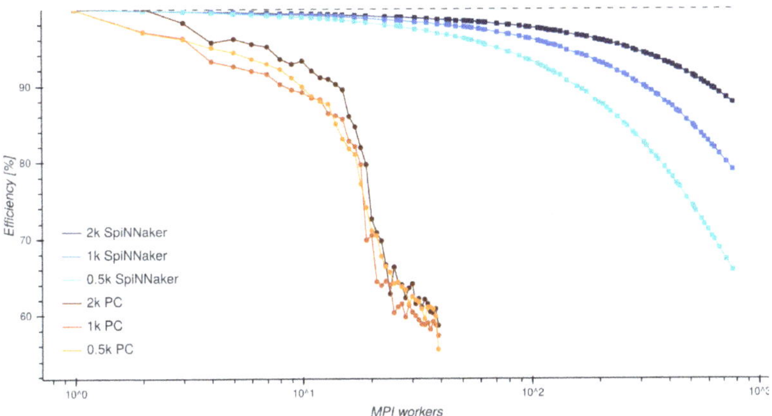

Figure 7. Comparison of Weak-scaling Efficiency for MPI-FED on a general purpose architecture and on SpiNNaker.

By contrast, the efficiency of the Server dips much faster, dropping below 90% as soon as the requested MPI workers outnumber the physical cores. It also remains fairly constant when changing the number of chunks. This appears reasonable as, for the high-speed CPU used in the test, the computation time is very small, but it suggests that other phases of the computation such as inter-process communication and thread management have a significant impact on the efficiency of the algorithm.

As a side-experiment, we evaluated the impact of the size of the *FED* buffer distributing data among the *MPI workers* on the measured scaling efficiency. Figure 8 shows the scaling efficiency of two experiments—the former distributes the *FED* chunks to be analyzed as 1000 256-Byte packets. The latter broadcasts the same amount of data, formatted as 125 2-kB packets. Figure 8 highlights that the two scaling efficiency tracks are comparable, meaning that the size of packets used to distribute *FED* chunks among the *MPI workers* does not impact the benchmark results for the general purpose architecture.

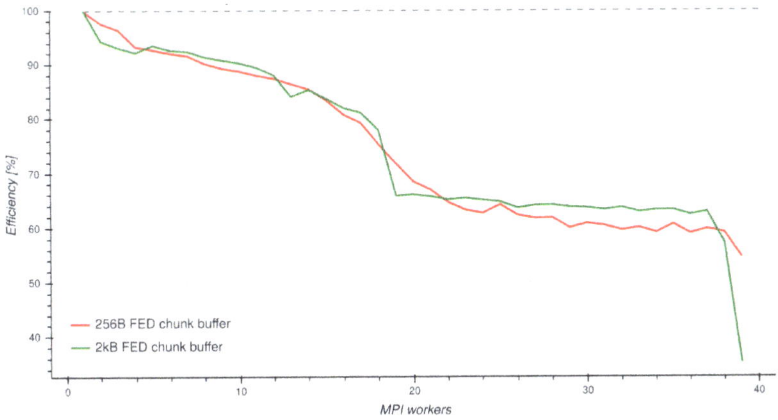

Figure 8. Efficiency of the general purpose architecture for different *FED* buffer sizes.

Finally, we can make a comparison of the power efficiency on the two architectures by using estimated consumption based on the nominal values from the CPU [30] and SpiNNaker [31] data-sheets. For the Intel Xeon, we consider the peak and idle powers at the values of $P_{peak} = 11,030$ mW and $P_{idle} = 6320$ mW, and we hypothesize that the number of active physical cores (out of the available 20), $f(x)$, can be expressed as a function of the active MPI workers x as $f(x) = ceil(\frac{x+1}{2})$. The appearance of the term $x+1$ rather than x is because there is one Controller process that has the task of distributing the data and patterns to the MPI workers. Based on this assumption, we assign a power consumption of P_{peak} to the active cores and of P_{idle} to every other core; thus the estimated power consumption with respect to the number of MPI workers x is $P(x) = P_{peak} \cdot f(x) + P_{idle} \cdot (20 - f(x))$.

On the other hand, for SpiNNaker we consider the values of Idle Power per Chip $C_{idle} = 360$ mW, Idle Power per Core $P_{idle} = 20$ mW, Peak Power per Core $P_{peak} = 55.56$ mW, and the Off-Chip-Link power, $P_{link} = 6.3$ mW. The power estimation for SpiNNaker depends on the MPI execution context, which can be described by a pair of values (p, k) where $p \in [1, 16]$ is the number of active processors per chip and $k \in [1, 48]$ is the number of active chips. The power estimation formula can be expressed as a function of the number of active processors and chips as $P(p, k) = k \cdot (C_{idle} + (P_{peak} - P_{idle}) \cdot (p + 1) + P_{link}) + (48 - k) \cdot C_{idle}$. Counting $p+1$ processors to include the Monitor Processor on each core. Then, the estimated power given the number of MPI workers x is $P(x) = P(p,k)|min_k[p \cdot k = x + 1]$ As in the CPU case, we count $x+1$ processes to include the Controller process.

Given the architectural difference between the SpiNNaker and CPU machines, it is necessary to outline a fair method to evaluate the efficiency of the algorithm's implementation. We define power efficiency as the energy consumed to align a single pattern to the reference, measured in units of mJ/pattern, as a function of the parallelisation effort of the given system, expressed as a percentage of the total resources. The maximum energy efficiency is obtained when all resources are in use, corresponding to a parallelisation effort of 100%. For SpiNNaker it is easy to assume that 100% utilisation occurs when all 768 cores are busy (i.e., at 767 MPI workers), corresponding to an average energy consumption of 37.3 mJ/pattern. For the CPU utilisation, we can either consider 100% utilisation to be the situation where all physical cores are active, or the one where all the virtual cores are active (20 physical + 20 virtual, providing 39 MPI workers). In the first case, the estimated average energy consumption is of 51 mJ/pattern, with an estimated power saving of 27% in favour of SpiNNaker. In the second case, the energy is 43 mJ/pattern, with SpiNNaker consuming 13% less.

5. Conclusions

In this work, we presented an implementation of an MPI-based DNA sequence matching algorithm for evaluating two critical aspects of using one of the more promising neuromorphic emerging technology. As the first point, we benchmarking the SpiNNaker many-core neuromorphic platform and its MPI support, showing that the scaling performances are kept linear when an increasing number of cores is used during the computation. As the second point, we demonstrated that by using the spinMPI library, which provides MPI support for SpiNNaker, we could easily port algorithm implemented for standard computers on the many-core neuromorphic platform.

The MPI standard exposes a programming model for the development of parallel applications in a distributed memory environment without knowledge of the interconnections between the computing units of the underlying architecture. The implementation of MPI for a specific architecture is therefore expected to implement the most suitable features in order to exploit the available resources and to synchronise the computing flow.

In the case of SpiNNaker, the implementation of MPI must deal with a resource limit both in terms of memory and computing power. However, it can take advantage of the technology offered by on-chip routers, obtaining efficient communication. SpinMPI is also in charge of managing communication between the *MPI Runtime* running on the host computer and the SpiNNaker cores; this is done by using the ACP protocol and memory entities. This software stack creates a simple working framework offering a universally known programming model capable of making the SpiNNaker architecture available for a wide range of applications.

We have succeeded in performing a benchmark of the SpiNNaker board by using a highly-parallel implementation of a DNA matching algorithm. Results show that the scalability of the SpiNNaker board reaches an ideal profile (98% of efficiency) when using more than 100 processors, a 90% efficiency using 600 processors, reaching 88% efficiency when all 767 application processors are used.

Author Contributions: Conceptualization, G.U. and F.B.; methodology, G.U., F.B. and A.A.; software, E.F., E.P. and F.B.; validation, E.F., E.P. and F.B.; formal analysis, E.F., E.P. and F.B.; investigation, G.U., E.F., E.P.; resources, G.U., A.A., and E.M.; data curation, E.P.; writing–original draft preparation, G.U., E.F., E.P. and F.B.; visualization, G.U., E.F. and E.P.; supervision, G.U.; project administration, G.U., A.A., and E.M.; funding acquisition, G.U. and E.M.

Funding: This research was funded by European Union Horizon 2020 Programme [H2020/2014-20] grant number 785907.

Acknowledgments: The research leading to these results has received funding from European Union Horizon 2020 Programme [H2020/2014-20] under grant agreement no. 785907 [HBP-SGA2].

Conflicts of Interest: The authors declare no conflict of interest.

Abbreviations

The following abbreviations are used in this manuscript:

SpiNNaker	Spiking Neural Network Architecture
Dynap-SEL	Dynamic Asynchronous Processor Scalable and Learning
BrainScaleS	Brain-inspired multiscale computation in neuromorphic hybrid systems
FED	Fast string matching method for Encoded DNA sequences

References

1. Mead, C. Neuromorphic electronic systems. *Proc. IEEE* **1990**, *78*, 1629–1636. [CrossRef]
2. Boahen, K.A. Point-to-point connectivity between neuromorphic chips using address events. *IEEE Trans. Circuits Syst. II Analog Digit. Signal Process.* **2000**, *47*, 416–434. [CrossRef]
3. Furber, S. To build a brain. *IEEE Spectr.* **2012**, *49*, 44–49. [CrossRef]
4. Liu, C.; Bellec, G.; Vogginger, B.; Kappel, D.; Partzsch, J.; Neumärker, F.; Höppner, S.; Maass, W.; Furber, S.B.; Legenstein, R.; et al. Memory-efficient deep learning on a SpiNNaker 2 prototype. *Front. Neurosci.* **2018**, *12*, 840. [CrossRef]
5. Blin, L.; Awan, A.J.; Heinis, T. Using Neuromorphic Hardware for the Scalable Execution of Massively Parallel, Communication-Intensive Algorithms. In Proceedings of the 2018 IEEE/ACM International Conference on Utility and Cloud Computing Companion (UCC Companion), Zurich, Switzerland, 17–20 December 2018; pp. 89–94. [CrossRef]
6. Sugiarto, I.; Liu, G.; Davidson, S.; Plana, L.A.; Furber, S.B. High performance computing on spinnaker neuromorphic platform: A case study for energy efficient image processing. In Proceedings of the 2016 IEEE 35th International Performance Computing and Communications Conference (IPCCC), Las Vegas, NV, USA, 9–11 December 2016; pp. 1–8. [CrossRef]
7. Barchi, F.; Urgese, G.; Macii, E.; Acquaviva, A. An Efficient MPI Implementation for Multi-Core Neuromorphic Platforms. In Proceedings of the 2017 New Generation of CAS (NGCAS), Genova, Genoa, 6–9 September 2017; pp. 273–276. [CrossRef]
8. Furber, S. Large-scale neuromorphic computing systems. *J. Neural Eng.* **2016**, *13*, 051001. [CrossRef]
9. Schuman, C.D.; Potok, T.E.; Patton, R.M.; Birdwell, J.D.; Dean, M.E.; Rose, G.S.; Plank, J.S. A survey of neuromorphic computing and neural networks in hardware. *arXiv* **2017**. arXiv:1705.06963.
10. Young, A.R.; Dean, M.E.; Plank, J.S.; Rose, G.S. A Review of Spiking Neuromorphic Hardware Communication Systems. *IEEE Access* **2019**. [CrossRef]

11. Schemmel, J.; Grübl, A.; Hartmann, S.; Kononov, A.; Mayr, C.; Meier, K.; Millner, S.; Partzsch, J.; Schiefer, S.; Scholze, S.; et al. Live demonstration: A scaled-down version of the brainscales wafer-scale neuromorphic system. In Proceedings of the 2012 IEEE International Symposium on Circuits and Systems, Seoul, Korea, 20–23 May 2012; p. 702. [CrossRef]
12. Moradi, S.; Qiao, N.; Stefanini, F.; Indiveri, G. A scalable multicore architecture with heterogeneous memory structures for dynamic neuromorphic asynchronous processors (dynaps). *IEEE Trans. Biomed. Circuits Syst.* **2017**, *12*, 106–122. [CrossRef]
13. Davies, M.; Srinivasa, N.; Lin, T.H.; Chinya, G.; Cao, Y.; Choday, S.H.; Dimou, G.; Joshi, P.; Imam, N.; Jain, S.; et al. Loihi: A neuromorphic manycore processor with on-chip learning. *IEEE Micro* **2018**, *38*, 82–99. [CrossRef]
14. Furber, S.B.; Galluppi, F.; Temple, S.; Plana, L. The spinnaker project. *Proc. IEEE* **2014**, *102*, 652–665. [CrossRef]
15. Furber, S.; Lester, D.; Plana, L.; Garside, J.; Painkras, E.; Temple, S.; Brown, A. Overview of the SpiNNaker System Architecture. *Comput. IEEE Trans.* **2013**, *62*, 2454–2467. [CrossRef]
16. Brown, A.D.; Furber, S.B.; Reeve, J.S.; Garside, J.D.; Dugan, K.J.; Plana, L.A.; Temple, S. SpiNNaker—Programming model. *IEEE Trans. Comput.* **2015**, *64*, 1769–1782. [CrossRef]
17. Urgese, G.; Barchi, F.; Macii, E. Top-down profiling of application specific many-core neuromorphic platforms. In Proceedings of the 2015 IEEE 9th International Symposium on Embedded Multicore/Many-core Systems-on-Chip, Turin, Italy, 23–25 September 2015; pp. 127–134. [CrossRef]
18. Urgese, G.; Barchi, F.; Macii, E.; Acquaviva, A. Optimizing network traffic for spiking neural network simulations on densely interconnected many-core neuromorphic platforms. *IEEE Trans. Emerg. Top. Comput.* **2018**, *6*, 317–329. [CrossRef]
19. Rowley, A.G.D.; Brenninkmeijer, C.; Davidson, S.; Fellows, D.; Gait, A.; Lester, D.; Plana, L.A.; Rhodes, O.; Stokes, A.; Furber, S.B. SpiNNTools: The execution engine for the SpiNNaker platform. *Front. Neurosci.* **2019**, *13*, 231. [CrossRef] [PubMed]
20. Rhodes, O.; Bogdan, P.A.; Brenninkmeijer, C.; Davidson, S.; Fellows, D.; Gait, A.; Lester, D.R.; Mikaitis, M.; Plana, L.A.; Rowley, A.G.; et al. sPyNNaker: A Software Package for Running PyNN Simulations on SpiNNaker. *Front. Neurosci.* **2018**, *12*. [CrossRef] [PubMed]
21. Barchi, F.; Urgese, G.; Siino, A.; Di Cataldo, S.; Macii, E.; Acquaviva, A. Flexible on-line reconfiguration of multi-core neuromorphic platforms. *IEEE Trans. Emerg. Top. Comput.* **2019**. [CrossRef]
22. Soni, K.K.; Vyas, R.; Sinhal, A. Importance of String Matching in Real World Problems. *Int. J. Eng. Comput. Sci.* **2014**, *3*, 6371–6375.
23. Boyer, R.S.; Moore, J.S. A fast string searching algorithm. *Commun. ACM* **1977**, *20*, 762–772. [CrossRef]
24. Horspool, R.N. Practical fast searching in strings. *Softw. Pract. Exp.* **1980**, *10*, 501–506. [CrossRef]
25. Reinert, K.; Dadi, T.H.; Ehrhardt, M.; Hauswedell, H.; Mehringer, S.; Rahn, R.; Kim, J.; Pockrandt, C.; Winkler, J.; Siragusa, E.; et al. The SeqAn C++ template library for efficient sequence analysis: A resource for programmers. *J. Biotechnol.* **2017**, *261*, 157–168. [CrossRef]
26. Kim, J.W.; Kim, E.; Park, K. Fast Matching Method for DNA Sequences. In *Combinatorics, Algorithms, Probabilistic and Experimental Methodologies*; Chen, B., Paterson, M., Zhang, G., Eds.; Springer: Berlin/Heidelberg, Germany, 2007; pp. 271–281.
27. Xue, Q.; Xie, J.; Shu, J.; Zhang, H.; Dai, D.; Wu, X.; Zhang, W. A parallel algorithm for DNA sequences alignment based on MPI. In Proceedings of the 2014 International Conference on Information Science, Electronics and Electrical Engineering, Sapporo, Japan, 26–28 April 2014; Volume 2, pp. 786–789.
28. Amdahl, G.M. Validity of the Single Processor Approach to Achieving Large Scale Computing Capabilities. In Proceedings of the AFIPS '67 Spring Joint Computer Conference, Atlantic, NJ, USA, 18–20 April 1967; pp. 483–485.
29. Gustafson, J.L. Reevaluating Amdahl's Law. *Commun. ACM* **1988**, *31*, 532–533. [CrossRef]

30. Intel Xeon Processor Scalable Family, Datasheet, Volume One: Electrical. 2018. Available online: https://www.intel.com/content/www/us/en/processors/xeon/scalable/xeon-scalable-datasheet-vol-1.html (accessed on 1 November 2019).
31. Painkras, E.; Plana, L.A.; Garside, J.; Temple, S.; Galluppi, F.; Patterson, C.; Lester, D.R.; Brown, A.D.; Furber, S.B. SpiNNaker: A 1-W 18-Core System-on-Chip for Massively-Parallel Neural Network Simulation. *IEEE J. Solid-State Circuits* **2013**, *48*, 1943–1953. doi:10.1109/JSSC.2013.2259038. [CrossRef]

MDPI
St. Alban-Anlage 66
4052 Basel
Switzerland
Tel. +41 61 683 77 34
Fax +41 61 302 89 18
www.mdpi.com

Electronics Editorial Office
E-mail: electronics@mdpi.com
www.mdpi.com/journal/electronics

www.ingramcontent.com/pod-product-compliance
Lightning Source LLC
LaVergne TN
LVHW070614170726
843515LV00004B/76

* 9 7 8 3 0 3 6 5 1 7 3 4 6 *